THE HAWAIIAN GOOSE

Frontispiece: Pair of adult Hawaiian Geese at The Wildfowl Trust's Martin Mere Reserve, Lancashire. Photo: Brian Gadsby.

THE HAWAIIAN GOOSE

An Experiment in Conservation

JANET KEAR and A. J. BERGER

T. & A. D. POYSER
Calton

First published 1980 by T & AD Poyser
Print-on-demand and digital editions published 2010 by T & AD Poyser, an imprint of A&C Black Publishers Ltd, 36 Soho Square, London W1D 3QY

www.acblack.com

ISBN (print) 978-1-4081-3758-1
ISBN (epub) 978-1-4081-3759-8
ISBN (e-pdf) 978-1-4081-3757-4

A CIP catalogue record for this book is available from the British Library

This is a print-on-demand edition produced from an original copy.

It is produced using paper that is made from wood grown in managed sustainable forests. It is natural, renewable and recyclable. The logging and manufacturing processes conform to the environmental regulations of the country of origin.

Printed and bound in Great Britain

FSC www.fsc.org MIX Paper from responsible sources FSC® C013604

Dedicated to
Sir Peter Scott and Ah Fat Lee

Contents

Acknowledgements 13

Introduction 15

1 THE HAWAIIAN ISLANDS 17

2 THE HAWAIIAN GOOSE OR NENE 35

3 THE NENE IN CAPTIVITY (especially in Hawaii) 59

4 THE NENE AT SLIMBRIDGE 79

5 THE RELEASE OF CAPTIVE-REARED NENE 100

6 THE EXPERIMENT IN CONSERVATION 115

Appendices

1 Endemic, extinct, rare and endangered Hawaiian birds 125

2 Weights of adult Nene 129

3 Measurements of Slimbridge adult Nene 130

4 The egg and clutch size of Nene 131

5 Gosling weights 133

6 Nene breeding performance in captivity 134

7 Mortality of Slimbridge Nene 138

8 Rainfall at Keauhou and Kahuku Sanctuaries 140

Bibliography 143

Index 151

List of Figures

1 Map of Pacific Basin 18

2 The Hawaiian chain of islands 19

3 Map showing presumed regions of origin of endemic Hawaiian birds 22

4 Head of Hawaiian Goose and of Atlantic Canada Goose 25

5 Map of Hawaii made in 1820s 27

6 Map of Hawaii 1944 29

7 Nene range prior to 1900 and after 1900 30

8 Sonagram comparison of adult Nene and Atlantic Canada Goose calls 36

9 Trachea of adult Nene 37

10 Weight variation of adult Nene at Slimbridge 38

11 Skull of adult Nene 39

12 Feet of Nene and Canada Goose 40

13 *Deschampsia nubigena*, an endemic Hawaiian grass 43

14 *Hypochaeris radicata* or gosmore 44

15 *Coprosma ernodeoides* or kukaenene 45

16 *Styphelia tameiameiae* or pukeawe 45

17 *Sonchus oleraceus*, sow thistle or pualele 46

18 *Physalis peruviana*, cape gooseberry or poha 47

19 *Vaccinium peleanum* or ohela 47

20 Down plumule from breast of female Nene 51

21 Sonagram of contact call of Nene gosling 53

22 Sonagram of sleepy call of Nene gosling

23 Sonagram of distress call of Nene gosling

24 Wings of adult and juvenile Nene

25 Production of Nene at Slimbridge in 1952

26 Nene deaths at Slimbridge plotted against age

27 Slimbridge Nene deaths in relation to month

28 Nene sperm

29 Number of eggs plotted against age of female – Slimbridge

30 Average egg weights plotted against age of female – Slimbridge

31 Fertile eggs plotted against age of gander – Slimbridge

32 Number of goslings plotted against age of female – Slimbridge

33 Incidence of atherosclerosis in dead Nene at Slimbridge

34 Growth curve of Nene goslings at Slimbridge

35 Average weight of male Nene at Slimbridge in winter

36 Locations of Keauhou and Kahuku Sanctuaries

37 Location of release site of Slimbridge Nene in Haleakala Crater

Tables

1 Average measurements of adult Nene at Slimbridge

2 Nene restoration project record

List of plates

Frontispiece: Pair of adult Hawaiian Geese, Martin Mere

between pages 48 and 49

1 Small inbred flock of Hawaiian Geese held by Herbert Shipman at Keaau, Hawaii, prior to 1949

2 Male Hawaiian Goose at Slimbridge

3 Heads of the four nearest relatives of the Hawaiian Goose

4 Small kipuka surrounded by an *aa* lava flow
Inconspicuousness of Nene on *aa* lava flow

5 Pair of Nene in Keauhou Sanctuary

6 Various intensities of 'upright threat' posture

7 Intensities of 'forward threat' posture

8 Small part of Nene rearing facilities at Pohakuloa, Hawaii
Moulting pair of Nene with young at Pohakuloa

9 Nene pens at Pohakuloa,
Ah Fat Lee with pair of Nene and four downy young

10 Gander standing guard while goose is incubating
David Woodside of Hawaii Division of Fish & Game at Nene nest in Keauhou Sanctuary

11 Nest covered with down feathers by goose before leaving
Wild Nene nest with four eggs

12 'Wigwam' constructed at Slimbridge providing wind-proof nesting site with gander on guard

13 Female Nene with week-old goslings at Slimbridge
Week-old Canada Goose gosling for comparison

14 Wild Hawaiian Goose with six-week-old young
Five-week-old goslings at Slimbridge

15 Immature Nene ready for release into the wild

16 Flock of Nene at Slimbridge

17 Crating the first shipment of Slimbridge-bred Hawaiian Geese

18 Typical habitat near release pen in Keauhou 2 Sanctuary
Difficult-to-locate Nene amongst introduced mullein in Keauhou 2 Sanctuary

19 David Woodside and Ernest Kosaka near release pen in Keauhou 2 Sanctuary
The release pen in Keauhou Sanctuary

20 Female Nene at Slimbridge in full wing moult

21 The preen gland

22 Normal Nene goslings compared with 'hairy' or 'cottony' down goslings

23 Lesions of possible avian pox on feet of Slimbridge Nene

24 Deeply furrowed neck feathers of Nene dibbling in water

Acknowledgements

We are especially indebted to Ah Fat Lee, Ernest Kosaka, Ronald L. Walker, and David H. Woodside of the Hawaii Division of Fish and Game for their kindness in providing photographs and unpublished data on the Nene propagation programme at Pohakuloa. Their splendid co-operation was essential in order to provide accurate information on many aspects of the project in Hawaii. Mr Woodside also generously served as a highly knowledgeable guide for A.J.B.'s visits to the release sites on Hawaii, and he read early versions of the manuscript for Chapters 1, 3, and 5. Mr Joseph Medeiros, District Wildlife Biologist on Maui, provided information on the release site in Haleakala Crater, and he read an early version of appropriate parts of the manuscripts. Mr Robert L. Barrel, State Director for Hawaii of the National Park Service, and Win Banko provided information on the Nene programmes for Volcanoes National Park and Haleakala National Park. We thank Dr Charles H. Lamoureux for his kindness in identifying various species of plants. Illustrations were drawn by Sir Peter Scott, Tim Halliday, Joe Blossom, Gwyn Williams, John Turner, and Sue Monden (who prepared the fine pen-and-ink drawings of favourite Nene foods from plant specimens collected on Hawaii). Dr Paul Johnsgard provided the sonagram of the adult call. Additional photographs were kindly lent by Lady Scott, E. E. Jackson, J. B. Blossom, P. Wallis, M. Brown, J. V. Beer, G. Williams, and B. Gadsby. Dr S. Dillon Ripley, Mr S. T. Johnstone, Mr M. R. Lubbock, Mr M. Ounsted, Mr T. W. Comer, and Dr Ray Erickson gave useful information on the egg-laying of Nene in their care. Simon Mulholland, John Turner, and Gwyn Williams helped prepare some of the text and appendices, and J. Gordon Read, Keeper of Archives at Liverpool Museum, assisted with information about the 13th Earl of Derby. Peter J. Morgan and Clem Fisher found the type specimen of the Nene for us. Sir Peter Scott, Professor G. V. T. Matthews, Mr Ralph Hodgson, Brigadier S. E. M. Goodall, Dr Myrfyn Owen and Mr

Andrew Dawnay kindly read the text and suggested many improvements. Mrs Sally Oshiro, Mrs J. Everett, Mrs E. Temple Carrington, and Mrs Lynda Seddon typed the manuscript.

Introduction

What is so special about the Hawaiian Goose that a book need be written about it? The answer is that it has become a classic symbol of Man's unconscious destruction of nature and, later, of his conscious effort to conserve. As is so tragically typical for many island birds, Man reduced the numbers of this unique, tame, land-living goose almost to extinction – from an estimated 25,000 in the 18th Century, to less than 50 by the 1940s. Usually nothing can be done when a point so low is reached, but in the case of the Nene (the name given to the bird by the Polynesians), international effort achieved a great deal – aviculture stepped in and saved the species. On the island itself pressure was applied to the government agency for funds to set up a propagation station at Pohakuloa in 1949, especially to breed the bird for release into the wild. There, the birds were placed in the devoted care of Ah Fat Lee. Abroad, interest in the goose resulted in non-governmental effort, largely encouraged by one man, Sir Peter Scott, who had established the famous Wildfowl Trust at Slimbridge, England, in 1946. There, the world's largest and most representative collection of waterfowl was built up. As important as the living birds, was Scott's determination to use them as a means of educating the public to the value of natural things, and as the basis for thorough-going research into the biological requirements of wild and captive waterfowl. Thus a happy set of circumstances provided the opportunity for a joint project between Hawaiian and British workers just at the time when the Nene population reached its lowest ebb. It is appropriate that this book can bring together the experience gained in both countries over the last quarter century.

1: The Hawaiian Islands

It was in 1778 that Captain James Cook named the Sandwich Islands in honour of John Montague, the 4th Earl of Sandwich.* Later, in recognition of the native inhabitants, they became known as the Hawaiian Islands. Anthropologists believe that the Polynesian ancestors of the Hawaiian people settled the islands between 500 and 750 AD. Millions of years before that time, however, there were no islands at all, and the Pacific Ocean rolled as an unbroken expanse of water for some 5,500 nautical miles between North America and Japan (Figure 1).

Then, a series of volcanoes erupted along a fracture zone in the floor of the ocean. They built up a massive underwater mountain range, and some of the peaks spewed their lava above the surface of the ocean. These islands form the Hawaiian chain, extending for a distance of 1,600 nautical miles from Kure Atoll in the northwest to the island of Hawaii in the southeast (Figure 2).

The island of Oahu, on which is the capital Honolulu, first appeared about 10 million years ago. The islands to the northwest are much older. Midway Atoll, for example, was partially eroded by wave action before Miocene times, or more than 20 million years ago. By contrast, the Big Island of Hawaii is less than one million years old, and two of its volcanic mountains have continued to erupt during this century. During some eruptions, lava still flows all the way to the shore, thereby adding to the 10,440 km^2 (4,030 square miles) that already make it larger than all of the other islands combined.

* The Earl of Sandwich (1718–1792) was a British First Lord of the Admiralty who has three claims to immortality. First, in having his servant place a piece of meat between two slices of bread – so that he could remain longer at the gaming table – he invented the sandwich. Secondly, he told Samuel Foote, the contemporary actor and dramatist: 'Foote, I have often wondered what catastrophe would bring *you* to your end; but I think you must either die of the pox, or the halter.' To which came the celebrated reply: '*That* will depend upon one of two contingencies – whether I embrace your lordship's mistress, or your lordship's principles.' And thirdly, many of the unique birds and plants of Hawaii, including the Nene, bear his name.

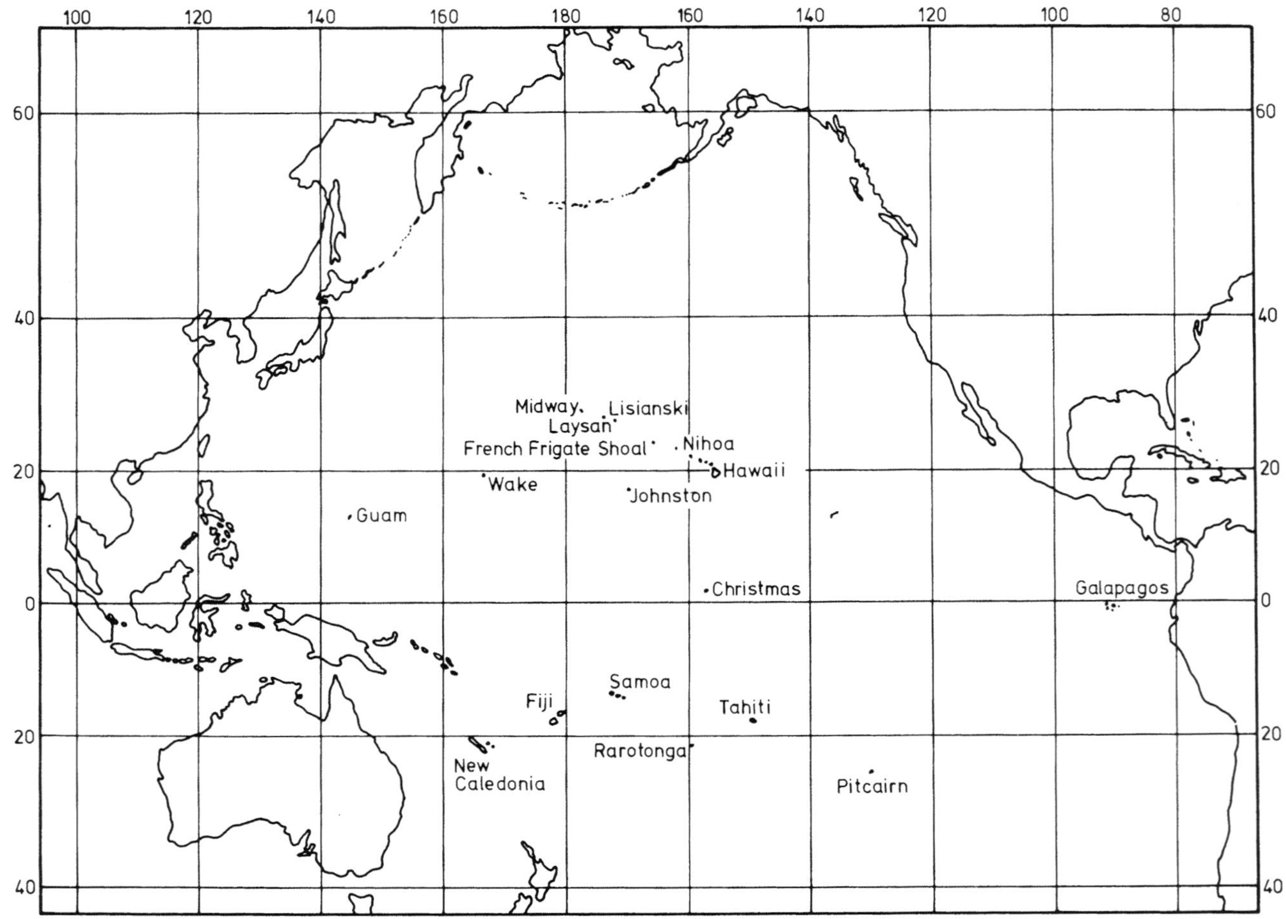

Fig. 1. Map of Pacific Basin, showing location of the Hawaiian Islands.

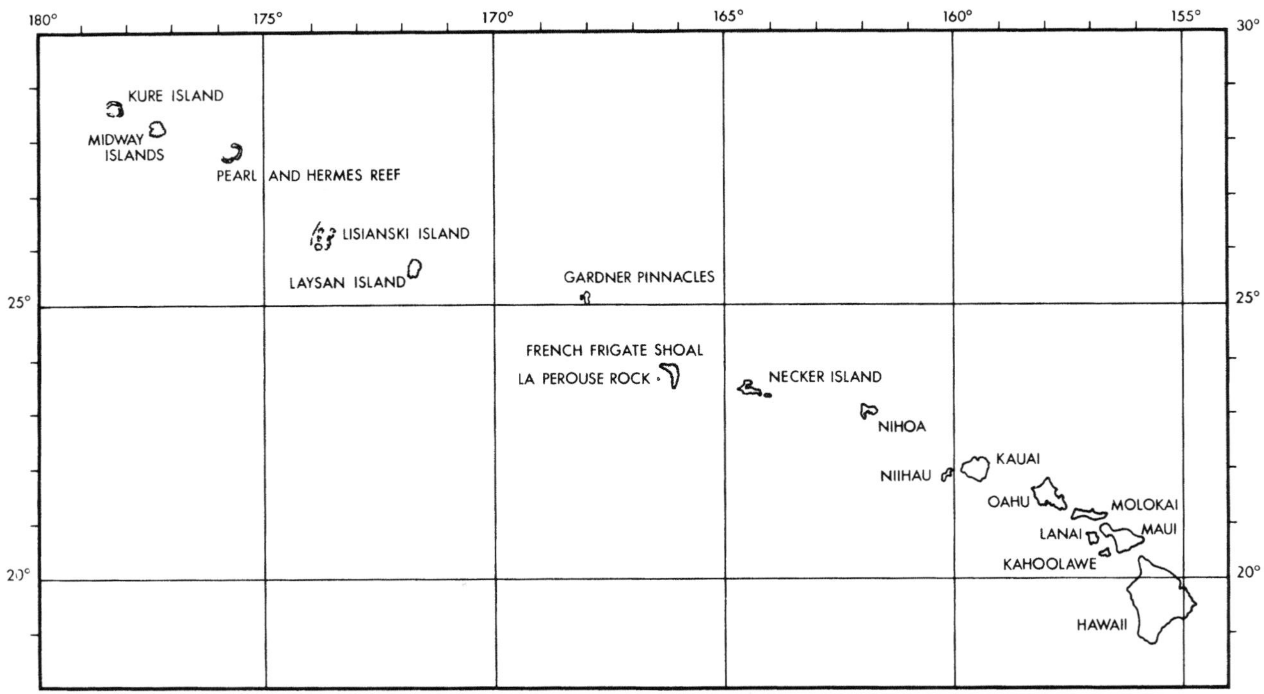

Fig. 2. The Hawaiian chain of islands. Reprinted from Andrew J. Berger, Hawaiian Birdlife, University Press of Hawaii, 1972 (19).

There are two distinct groups of islands in the chain. One consists of the very old, low, flat islands, reefs, and atolls formerly known as the Leeward Islands, but which the National Board of Geographic Names now calls the Northwestern Hawaiian Islands. These extend from Nihoa Island to Kure Atoll, a distance of about 1,930 km (1,200 miles). All are very small – the largest, Laysan Island, contains about 481 ha (1,020 acres) and Nihoa has but 63 ha (156 acres). Nihoa and Necker Islands, Gardner Pinnacles, and La Perouse Rock are formed of ancient lava rock. The others, however, now consist only of low sand and coral islands with no visible evidence of their volcanic bases.

The Hawaiian Islands National Wildlife Refuge includes all of the islands from Nihoa to Pearl and Hermes Reef. This refuge, established by President Theodore Roosevelt in 1909, provides nesting habitat for millions of seabirds, for the rare Laysan Duck *Anas platyrhynchos laysanensis*, for two species of endemic land birds, and for the Green Sea Turtle *Chelonia mydas*. The Hawaiian Monk Seal *Monachus schauinslandi* also breeds on the sandy islands.

The Southeastern or Windward Hawaiian Islands are the 'high' islands, most of which are inhabited by man. These include the group that extends from Kauai and Niihau in the northwest to Hawaii in the southeast, and it is here that the Hawaiian Goose *Branta sandvicensis* is found.

THE ORIGIN OF HAWAIIAN BIRDS

There would seem to be little doubt that the first birds to reach the newly formed Hawaiian Islands were seabirds, which spend the non-breeding season flying over the trackless expanses of the open ocean. Eventually, however, land birds found their way, presumably blown by storms. A pair or small flock of birds could establish a colonising population if the island they reached provided food, shelter, and safe breeding places.

Between 1778, when Captain Cook 'discovered' the Hawaiian Islands, and 1923, ornithologists had found 69 different kinds of birds (including three subspecies of seabirds) that were unique, or endemic, to the Hawaiian Islands. They did not have 69 different ancestors, however. Some of the ancestral populations gave rise to several species by adaptive radiation or evolutionary divergence into the many unoccupied habitats where food and other necessities were more than adequate to permit their offspring to thrive. Some of the random changes (mutations) in the genes of the birds resulted in the development of different kinds of bills and tongues, or other anatomical features that particularly suited certain ecological niches. So, over long periods of time, there evolved populations, or species, of birds that were unlike their ancestors and that bred only with their own kind. Most ornithologists believe that the 66 endemic land birds evolved from no more than 15 ancestral species. Nevertheless, given the age of the Hawaiian Islands, a successful colonisation needed to occur only once every 300,000 years.

The ancestors of the Hawaiian land birds are thought to have reached the islands from the regions shown in Figure 3.

THE ENDEMIC HAWAIIAN BIRDS

Twelve families of birds (if we include seabirds) have endemic genera, species, or subspecies in Hawaii, and one entire family, the Hawaiian Honeycreepers, is unique to the islands. A tabulation of these unique Hawaiian birds is given in Appendix 1.

Unfortunately, Hawaii has one of the most tragic histories in the world for the destruction of its endemic animals and plants. Of the unique birds, more than a third are believed to be extinct, and, of the remaining forms, nearly 40% are now considered to be rare and in danger of becoming extinct in the not distant future. Among these rare species are the Hawaiian Goose or Nene (pronounced 'nay-nay'), the Hawaiian Duck or Koloa *Anas platyrhynchos wyvilliana*, and all of the pond, marsh, and stream birds.

THE NENE

The earliest preserved skin of the Nene taken directly from the wild, is now in the United States National Museum, Washington DC, and was collected in 1840 during the visit to Hawaii by the US Exploring Expedition (103). Nicholas Vigors, however, had described and given the goose its scientific name *Bernicla sandvicensis* in 1833 at a meeting of the Zoological Society of London. The bird was living in the Society's gardens in Regent's Park at the time (3), and when it died, its skin (the type specimen) passed to the collection of T. C. Eyton, and later to that of Canon H. B. Tristram. Tristram's collection was purchased by the Liverpool Free Public Museums in 1896, and what remains of that original goose from the London Zoo is in Liverpool still.

Alexander Wetmore (131) described a fossil goose from Hawaii whose bones were found at a depth of about 24 m (75–80 feet), overlain by prehistoric lava on which a dense forest had grown. He

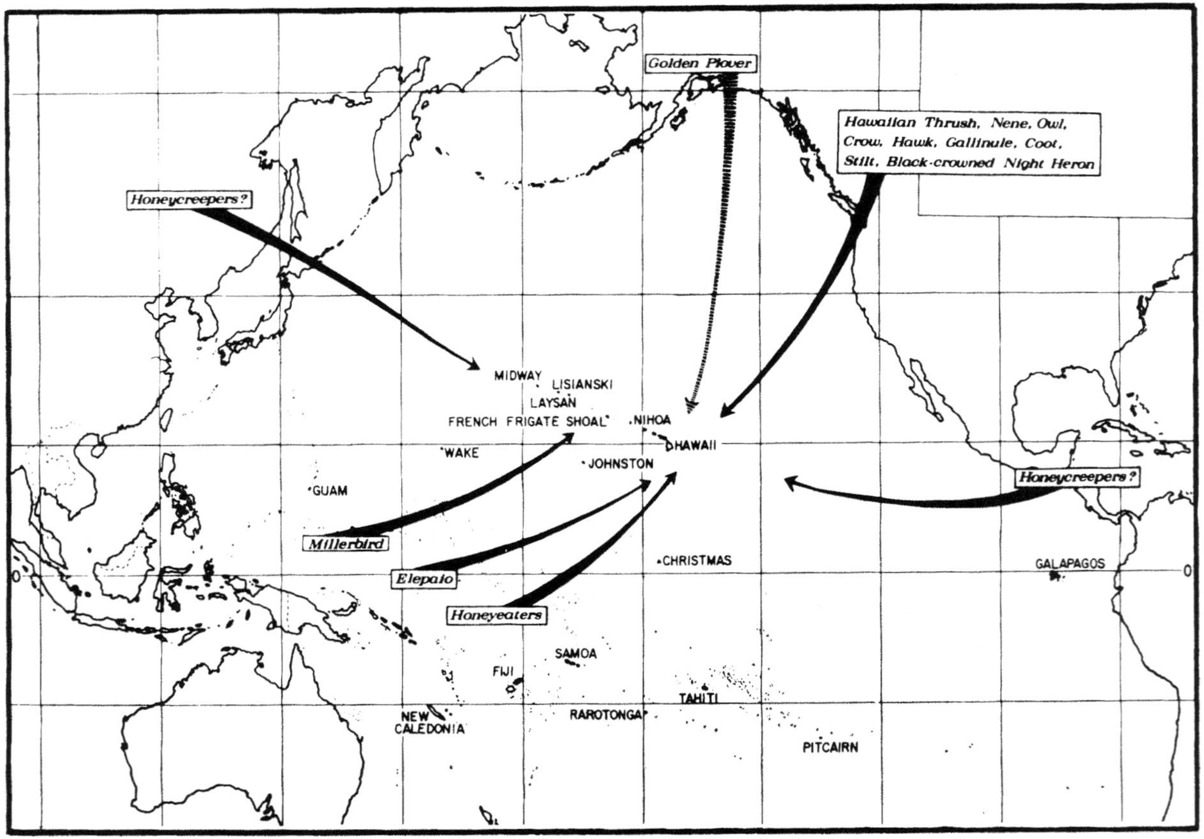

Fig. 3. *Map of the Pacific Basin to show regions from which the ancestors of endemic Hawaiian birds are presumed to have originated. The broken arrow indicates the annual migratory flights of the Pacific Golden Plover* (Pluvialis dominica fulva) *between Alaska and the Hawaiian Islands.* Reprinted from Berger, 1970 (17).

wrote that 'it would appear that the bones may belong to the Recent period, though it seems that they may have an age dating back thousands of years.' Wetmore named the new genus and species *Geochen rhuax*, and he remarked that 'this bird appears to have been as large as the living Cape Barren Goose (*Cereopsis novaehollandiae*) of southern Australia and to have been more similar to that bird than to any other now known. . . . *Geochen* shows no close alliance with the living Nene (*Nesochen sandvicensis*), except that both belong to the family Anatidae.'

Bones of another fossil goose were found on the island of Molokai during 1971 and 1972. This ancient goose was much larger than the Nene, and had a flat, rather than a keeled, sternum, reduced wing bones and large feet, suggesting that it was flightless. These bones are thought to be 25,000 years old (122).

Ornithologists believe that the Canada Goose *Branta canadensis* (Figure 4) is the closest living relative of the Nene and, therefore, that its ancestors came from North America. In fact, three species of North American duck are regular winter residents in the Hawaiian Islands, and the Cackling Canada Goose *B.c.minima* and 24 other waterfowl species have been reported one or more times as stragglers or chance migrants to Hawaii during the past century (19).

The Nene is certainly a native species of the island of Hawaii. Without giving any details, Brigham (31) in 1909 remarked that this goose was found 'only on the highlands of Hawaii and Maui.' It seems probable, however, that he had very little experience with it in the field. Wilson (133) stated: 'I heard that it nested in the crater of Haleakala [House of the Sun] on Maui, but I did not visit that place.' W. A. Bryan (35), Curator of Ornithology at the Bishop Museum, mapped the Nene in 1901 distributed only on Hawaii. Henshaw (62), who lived in Hawaii for ten years, stated in 1902 that 'the nene is now found chiefly, if not wholly, upon the island of Hawaii, although it is said to have nested in the past times in the crater of Haleakala on the island of Maui and, occasionally, to have been seen off the islands of Kauai and Niihau. At the present time, however, there is no reason to believe that the nene is found upon Maui, inquiry in 1891 failing to disclose that it has been seen there for several years. Its occurrence upon Kauai and Niihau was probably in the nature of an accident, if indeed it was not mistaken for one of the species of American Geese.' Perkins (104) wrote in 1903 that the Nene 'is said to have formerly frequented Haleakala, Maui, although neither in my many visits to the summit, nor when camping in the bottom of the great crater did I myself get sight of one. From the fact that parts of the crater as well as much of the higher slopes of Haleakala are extremely similar to some of its favourite haunts on Hawaii, and produce also some of its favourite food in abundance, its scarcity on Maui at the present time (if indeed it still exists there at all) is remarkable.' Pratt (107) cited the presence of Puunene [Nene hill] on the slopes of Haleakala as proof that the Nene once occupied Maui.

It is interesting to note that Titian Peale wrote in 1848 (103) that he observed the Nene 'on the volcanic mounts of the Island of Hawaii; they were generally in pairs at the season of our visit in the month of November, rarely four or five were seen together feeding on the berries of a very abundant species of Vaccinium growing on the old beds of lava; on these they become very fat, and were delicious eating; grass appeared, however, to be their ordinary food. We never saw them near water, which is scarce in those regions, our party being obliged to carry the necessary supplies for the journey in calabashes . . . What is most remarkable is the story related to us by natives, and which we have every reason to believe is a fact, that this Goose, which has the powers of flight which would enable it to move to as great distances as others of the genus, is limited to the single Island of Hawaii; rarely visiting any other islands of the group, although several are in sight.'

Baldwin (13) summarised his 1945 study of the history of the Nene by stating that 'complete proof of the former existence of the Nene on Maui is lacking, for no specimen is known to have been saved. Enough records are available, however, to indicate satisfactorily that a wild goose, presumably identical with the Nene, occurred on Maui. No specific breeding records are available.' In any event, the Nene has the smallest range of any species of goose.

THE NENE AND THE HAWAIIANS

Although several species of birds played an important role in Hawaiian mythology and culture, the Nene seems not to have been one of them. David Malo, an Hawaiian born on the island of Hawaii about 1793, wrote about Hawaiian history and culture. Malo's book was translated from Hawaiian into English by Nathaniel B. Emerson in 1898 and published by the Bernice P. Bishop Museum in 1951 (94). Malo wrote: 'The *nene*, which differs from all other birds, is of the size of the [muscovy] duck, has spotted feathers, long legs and a long neck. In its moulting season, when it comes down from the mountains, is the time when the bird-catchers try to capture it in the

Fig. 4. Head of the Hawaiian Goose (left) and the Atlantic Canada Goose (right). Tim Halliday.

uplands, the motive being to obtain the feathers, which are greatly valued for making *kahili*. Its body is excellent eating.' It is clear, however, that he had little or no personal contact with endemic birds, and especially the forest birds, as there are many errors in his descriptions. Brigham (29; 30; 32) does not mention the Nene in his extensive study of Hawaiian feather work. Nevertheless, several of the feather capes and tippets painted by Sarah Stone between approximately 1779 and 1785 (54) show brownish feathers, and it is possible that some of these were Nene feathers.

Of his experiences while visiting Kilauea Volcano on the island of Hawaii, Boddam-Whetham (24) wrote in 1876 that 'on returning to the [volcano] house, I found a very tempting repast ready, and amongst other luxuries was a strawberry-fed goose which had been enveloped in leaves and baked in a hole in the heated earth'* The Hawaiian wild geese frequent these mountains in great numbers, and are sure to be found feeding on the wild strawberries which abound in this neighbourhood. The natives declare that there is a lake somewhere on the mountains to which these celebrated geese retire, but it has never been discovered.'

Scott Wilson (133) also discovered that 'the flesh of this goose is good eating, and from it may be made the most excellent soup, which I remember to have formed the most delicious item amongst many other delicacies – as roasted goat, golden plover on toast, quail, bananas, bread-fruit, pineapples, custard apples, mangos – of my Christmas dinner at Kiholo on Hawaii.'

Henshaw (62) in 1902 noted that 'it is when leading their young that the old birds undergo the moult, and when deprived of their wing feathers and unable to fly, they, and the young, are easily run down by the fleet-footed natives and secured.' Wilson also remarked that 'the weird cry of the Nene . . . is very distinct from that of any other species that I know; and in olden times the bird was kept in captivity by the natives, acting as a sentinel by giving loud warning of the advent of a stranger.'

THE DECLINE OF THE NENE

Baldwin (13) presented a fine discussion of the historical record of the Nene from 1789 to May 1944, by which date he believed the total wild population to number about 50 birds. Baldwin's paper is of

* Volcano House, the hotel on the edge of the crater, was built as a grass house in 1846 when visitors were charged $1 a day, meals extra. Mark Twain stayed there in 1886; it is possible to do so today, and to see wild Nene not far away.

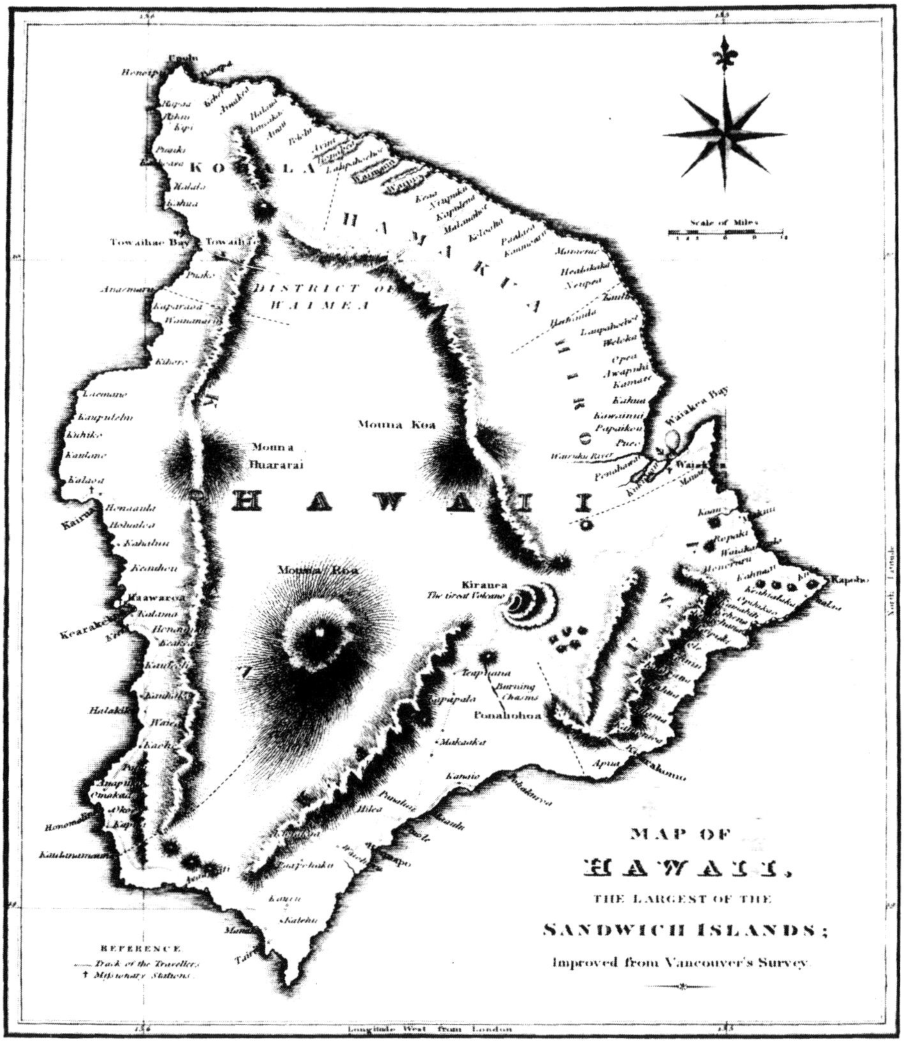

Fig. 5. *A map of the island of Hawaii made in the 1820s.* From Ellis, 1825 (51).

special value because it includes information obtained in conversations with long-time residents of Hawaii.

In his journal of the voyages of Captain Cook, William Ellis (50), the Assistant Surgeon, recorded that, at their arrival at Kealakekua Bay, Hawaii, 'the natives brought off several geese, which were quite tame; they are not unlike the Chinese geese; they called them Na-Na. By what means they procured them, we could not learn.'

In 1789, the King of Hawaii Island sent 12 birds to Captain John Mears of the ship *Iphigenia*, then anchored offshore from Kawaihae near the northern part of the leeward coast of the island.

There seems to be no mention of the Nene again until 1823, when William Ellis (52), the early missionary, wrote of his trip to Kilauea Volcano: 'On our way to the sulphur banks, we saw two flocks of wild geese, which came down from the mountains, and settled among the ohelo bushes near the pools of water. They were smaller than the common goose, had brown necks, and their wings were tipped with the same colour. The natives informed us there were vast flocks in the interior, although they were never seen near the sea' (Figure 5).

Of a trip to the dormant crater of Mt Hualalai in north Kona during July, 1864, Brigham (31) wrote that 'our guide shot two of the native geese . . . which were fine eating. The number of these geese has been much underrated. Although they are found only on the highlands of Hawaii and Maui, their number admits of the annual slaughter of several hundred without sensible diminution.' However, Brigham did not spend extended periods of time in Nene habitat, so that his comments should be interpreted with caution. Nevertheless, his statement does imply, as Baldwin pointed out, that 'their number in less remote areas was no longer large, though they were still abundant in the interior.'

Baldwin (13) suggested that 'a reasonable approximation' of the number of Nene on Hawaii during the late 18th century would be 25,000 birds. He summarised the decline as follows: 'Earliest available records indicate that a large population of Nene existed at least until 1823. Later ones show a decline, for in 1864, Brigham, although speaking optimistically, implied that the population was becoming smaller, as did MacFarlane in 1887. After 1890 we learn they are gone from Maui and scarce on Hawaii. From 1900 on the Nene has been rare.' He added that, as of 1944, 'the present range of the Hawaiian Goose, or Nene, incircles Mauna Loa [Long Mountain] on Hawaii between roughly 5,000 and 7,500 feet elevation and extends to or near the seashore in parts of Kau, North Kona, and South Kohala. Nene range was formerly more extensive on Hawaii than it now is, having decreased from 2,475 to 1,150 square miles. The Nene range of 1800 included larger tracts of lowland areas than that of 1900 to 1940. The present centre of the Nene population is Puuwaawaa north of Mt Hualalai. . . . Since 1930 the Nene popula-

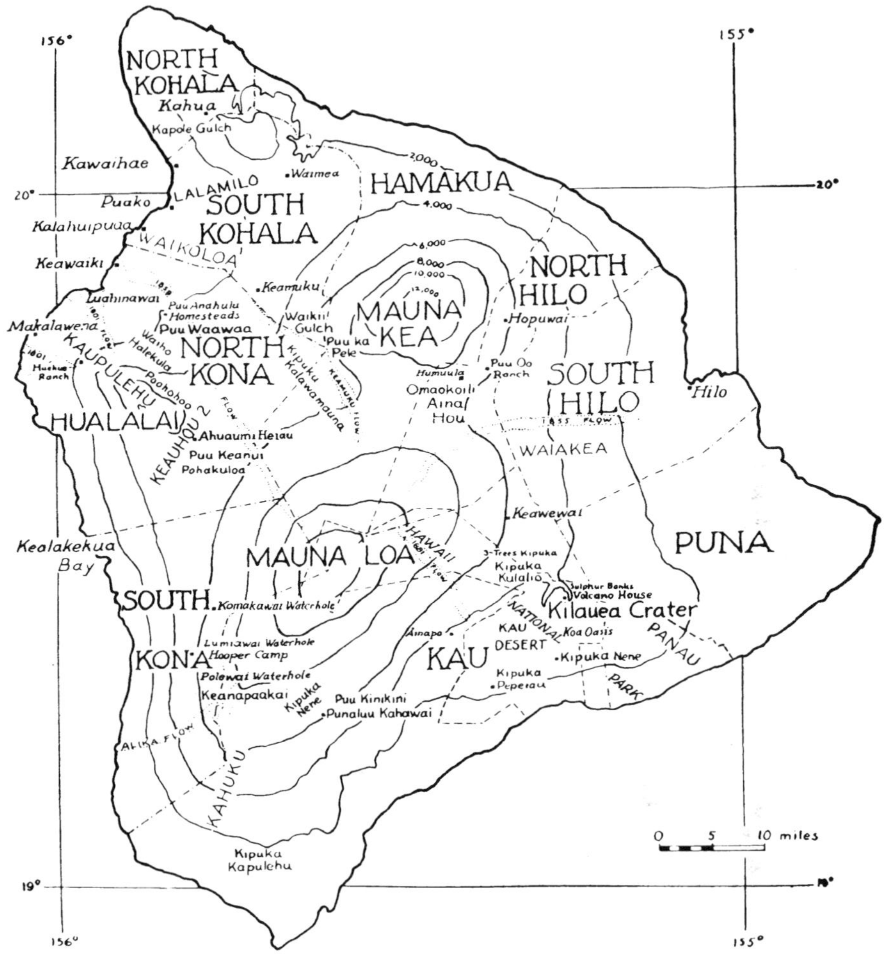

Fig. 6. A 1944 map of Hawaii. Reprinted, by permission, from Baldwin, 1945 (13).

tion appears to have remained constant, although the available evidence may be deceptive. The continuity of the wild Nene population has not yet been broken. The Nene has survived during the past 15 years at a low population level' (Figures 6 and 7).

The evidence *was* deceptive, and Charles and Elizabeth Schwartz, who conducted an 18-month game bird study in Hawaii from February 1946 through July 1947, saw no wild geese during their field work. They wrote (114) that 'this wildfowl is the next

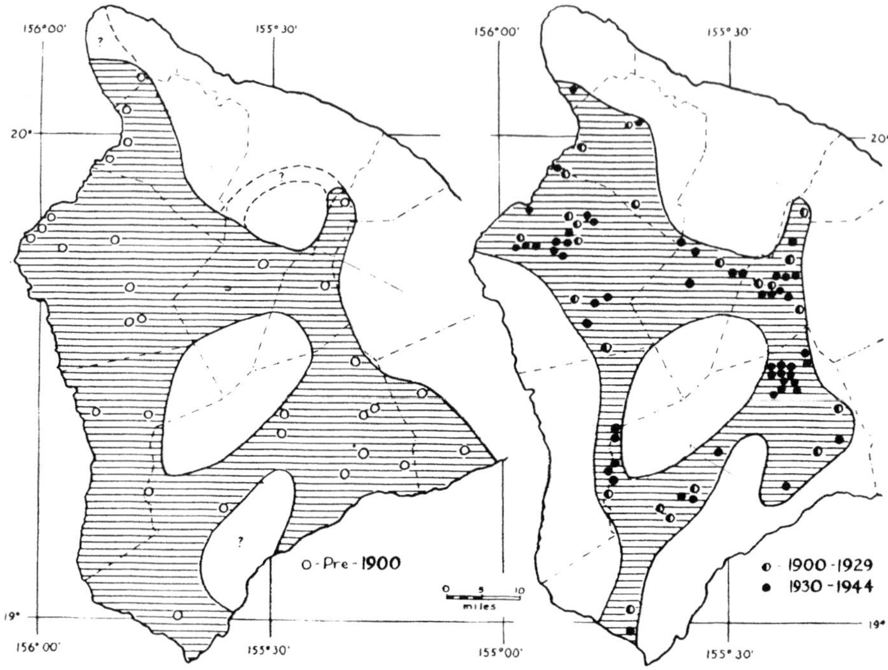

Fig. 7. Nene range prior to 1900 and after 1900. Reprinted, by permission, from Baldwin, 1945 (13).

Hawaiian, if not world, species facing imminent extinction. To permit this tragedy to occur without exerting more effort than has been to date is unpardonable.'

Moreover, by 1951, the total wild population was estimated by J. Donald Smith (120) to be no more than 30 birds. He wrote: 'Time for the Hawaiian Goose seems to be rapidly running out. It is appropriate, therefore, to emphasize that the Hawaiian Goose population today is even less than it was in 1944 and in 1947 when Baldwin and the Schwartzes made their field studies. Within the last three years, Nene have failed to return to certain locations that until recently were apparently highly preferred parts of their habitat. In 1944 Nene could be found consistently on the Puu Waawaa Ranch on the slopes of Mt Hualalai on Hawaii. They nested there, and young were seen frequently. The birds have now vanished from this specific locality and also from the entire Hualalai-North Kona region, all within the past several years. Although Nene certainly were not common, ranch hands during their fence patrol work usually saw at least a few geese in the years prior to 1940; however, in the last three years these same persons covering the same routes in the same manner have seen no geese.

'In the Kau Territorial Forest Reserve on Mauna Loa, there is an area to an elevation of 6,500 feet dotted with small, shallow, permanent pools and covered with an open ohia (*Metrosideros collina*) forest with a ground cover of lush grass and sedges. On maps of Hawaii, this area is labelled Kipuka Nene, a name given by the ancient Hawaiians testifying to its habitation by Nene for centuries. [Kipuka means an isolated patch of vegetation surrounded by recent lava flows (Plate 4).] In former years, the forest rangers frequently saw geese in this location during patrols in the winter months. In 1948, a ranger saw a flock of 17 Nene, and in January 1949, three were observed. Since then, no Nene has been seen, yet the ranger has continued to make his monthly patrols in the same routine manner . . .

'While it is true that no systematic count of the Hawaiian Goose has been made, on the basis of a continued constriction of its range since 1949 and the drop in sight report frequency, the population of wild Nene today can hardly exceed 30 birds. The startling drop in its numbers during the past few years probably is an indication that unless something is done to aid the geese, the existence of the species in a wild state will soon be only a memory.'

CAUSES OF THE DECLINE

There were undoubtedly multiple causes for the decline in the Nene population, and virtually all of them resulted from the settling in Hawaii of foreigners from many lands. Baldwin (13) summarised some of the causes:

'Activities of man: exploration, hunting with firearms, probable increase in capture of live birds and eggs, flushing and frightening birds from nests and foraging grounds, sandalwood gathering in uplands, ranching developments and activities, and building of beach resort homes and of military roads in uplands.

'Indirect agents of the white man's activities: introduced animals such as the rat, goat, sheep, cattle, horse, pig (new stock), ass, feral dog and cat, mongoose, game birds (pheasant, quail, guinea hen, jungle fowl, turkey, peafowl), the mynah, and the ant. Introduced plants such as pasture weeds and grasses, mesquite, thimble-berry, and pampas grass.

'Most of these factors have been adverse, but a few of the plants have provided preferred foods. Whatever the proportionate individual importance, the aggregate effect caused the drastic reduction of the Nene population.'

Cattle, horses, sheep, goats, and English pigs were set loose on the islands in the short period between 1778 and 1803. Moreover, when Captain Vancouver brought his second load of cattle to the island of Hawaii in 1794, King Kamehameha proclaimed a ten-year kapu (taboo) on the killing of the cattle and other introduced mammals, which prevented the common people from hunting them. The cattle, as well as sheep, were driven from Kealakekua Bay to the Waimea Plain, 'a great tract of luxuriant vegetation', where the animals were allowed to roam unrestrained. Vancouver wrote that observance of the ten-year kapu 'cannot fail to render the extirpation of these animals a task not easily to be accomplished.' He was, of course, correct.

These feral animals multiplied rapidly in the absence of their usual diseases and predators in a land with a year-around salubrious climate. As the populations increased, the mammals moved further into the virgin forests, slowly destroying many of them by trampling ground cover and by feeding on the seedlings and understorey plants. The years from 1815 to 1921 were later to be characterised as the 'cattle period in Hawaiian forestry.' L. A. Bryan (34) reported that more than 10,000 introduced mammals (such as cattle, goats, and pigs) were killed *every year* from 1921 to 1946 in the Forest Reserves alone on the island of Hawaii. The last feral cattle and horses were not exterminated from Mauna Kea [White Mountain] until the 1930s, and feral cattle are still found on some of the islands, including Hawaii (124).

These grazing, browsing, and rooting animals did not confine their activities to the rain forests, however, but also moved into the drier areas, including the gentle slopes of the mountains to elevations greater than 3,000 m (10,000 feet). So they roamed over nearly the entire extent of the range of the Nene.

Feral goats have been especially destructive to the native vegetation, and they find all of their requirements in the kipukas and on the sparsely vegetated lava flows inhabited by the Nene. Nearly 184,000 goat skins were exported from Hawaii in the period from 1885 to 1900 (141). Goats have also long been hunted for food, but they still plague the land. Rangers estimated in 1971 a population of 14,000 goats in Volcanoes National Park, Hawaii, which includes part of the Nene's present breeding range.

Efforts are being made to control the numbers of feral pigs in the Park, but there still exists a great overpopulation in many areas on Hawaii, and they are known to prey on the eggs and young of the Nene.

The mongoose was introduced to the Hamakua Coast of Hawaii in 1883 – in hopes that it would control the rats that were doing great damage in the sugarcane fields. The mongoose neither solved the rat problem nor remained in the sugarcane fields. For many years now, the mongoose has been found over most of the islands of Hawaii and Maui, from sea level to tree line and in the very wet (7,620 mm of annual rainfall) and the very dry (less than 500 mm of rain per year) areas. The mongoose is a likely predator of any ground nesting bird.

The precise role of feral cats and dogs as predators on the Nene has not been determined, although there is little doubt that they each sometimes prey on eggs, young birds, and flightless adults. The Hawaiians brought dogs to Hawaii with them, primarily as a source of food. Many other breeds of dog found their way to Hawaii after Captain Cook, and Kramer (85) wrote that 'feral or "loose" dog packs run free from time to time on all islands.' He added that 65 lambs were killed in one night on the Parker Ranch on Mauna Kea in 1961. Feral cats also are still fairly common on all of the islands.

There seems to be little doubt that the extinction of some of the endemic forest birds (collected for their colourful feathers) was hastened after shotguns were brought to Hawaii, and it seems reasonable to assume that many more Nene were killed when guns became common. Henshaw in 1902 (62) called attention to two aspects of Nene hunting when he wrote that 'the districts frequented by the bird most of the year, though remote and inaccessible, are now often visited by sportsmen, and the nene is rapidly diminishing in numbers. The time will inevitably come, and that soon, when this goose will need protection from sportsmen to save it from its otherwise inevitable fate of extermination.

'In this connection attention may well be called to the fact that, as the present law stands, the open months (from September 15 to February 1), when the killing of this goose is permissible, are almost precisely the ones in which it rears its young. The law, doubtless through a misapprehension of the facts, protects the bird when it least needs protection, and exposes it to slaughter when it is laying its eggs and leading about its young.'

The publication of unjustifiably optimistic statements on the status of the Nene undoubtedly has not been in the best interests of the species. Blackman wrote in 1944 (23), for example, that 'the

Hawaiian Goose, or *Nene*, once threatened with extinction, now has a better chance to survive on the slopes of the highest island mountain in the world.' The only reason for this optimistic view was that the practice of winter hunting 'was discontinued when protective measures were introduced in 1907, and although at the present time the Hawaiian Goose is by no means common, it now has a better chance to survive.' Seven years later the total wild population was estimated to be no more than 30 birds.

2: The Hawaiian Goose or Nene

The Nene stands about 50 cm (20 in) high, and weighs some 2 kg ($4\frac{1}{2}$ lb), making it the largest of Hawaii's native land birds. It is an attractive, medium sized goose, with plumage softly barred with fawn, brown and black, brown eyes, rather long black legs, black feet with reduced webbing and a black bill. The crown, the face, and a wide band down the back of the neck, are black and the underside of the tail is white. Cheeks, throat, and neck at front and sides are buff, and there is a narrow ring of black feathers encircling the base of the neck (39; 62; 111; 133). The neck feathering is more deeply furrowed than in other goose species (Figure 4).

How do ornithologists decide that the Nene is a goose? Popularly, the name is given to a variety of waterfowl species, but scientifically it is restricted to 'true geese' belonging to two genera: *Anser* and *Branta*. They are grazing birds that inhabit the northern hemisphere; sexes are similar, they tend to pair for life, and reach maturity in their second year. Their closest relatives are the swans, which may not look particularly like them, but whose anatomy and behaviour, including their well-developed and prolonged family relationships, indicate the link. The five species of *Branta* are known as the black geese and differ from the nine *Anser*, or grey geese, in having darker plumage and black legs and feet. Plate 3 shows the other black geese which are the Nene's nearest relatives, the closest probably being the Canada Goose.

VOICE

All true geese are rather noisy, and the Nene has a varied vocabulary in which two types of call are obvious. A low murmuring 'Nay-nay' is given with the bill closed. This is the sound commonly heard throughout the year between contented pairs and feeding flocks, and it is presumably from this that the Hawaiian Goose derives its Polynesian name.

As the breeding season approaches, sounds become louder and

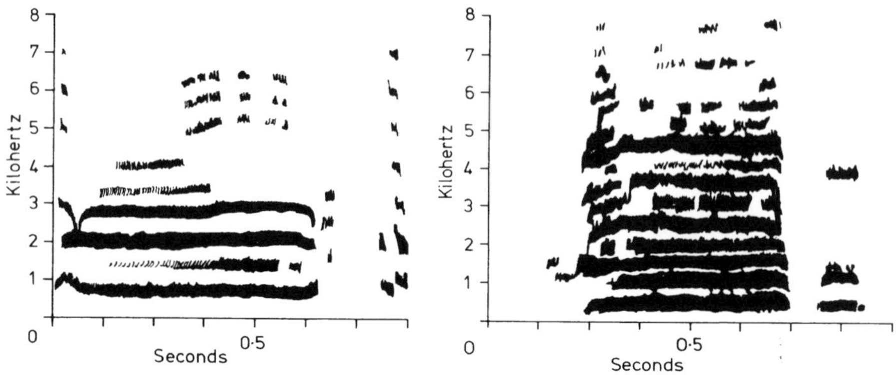

Fig. 8. Tracing of sonagram comparison of call of adult Nene (left) with adult Atlantic Canada Goose (from Johnsgard (68)).

more strident. Pairs, especially males, call with open bills and outstretched necks. They give shrill, slightly double syllabled cries, often followed by a series of shorter, more staccato notes and ending in a painful-sounding moan. The voice of the breeding Hawaiian Goose is not melodious at close quarters, but high in the lava fields, where pairs are crying their territorial challenges, it acquires quite a different quality.

Johnsgard (68) compared the call of the adult Nene with that of the Atlantic Canada Goose after he had changed the sounds into visual signals on a sonagram (Figure 8). The signals look very similar, with a basic, fundamental frequency and, at regular intervals above, a series of harmonics. In general, the longer the windpipe of the bird, the lower the fundamental frequency, and the greater the number of resonating harmonics. So the call of the small, short-necked Nene has a higher basic frequency with fewer harmonics than that of the Canada Goose. In simple terms, this means that its voice is shriller and does not carry quite so far.

The trachea or windpipe of the Nene was examined by P. S. Humphrey (63) and is illustrated in Figure 9. He found that it and its musculature were similar to other black geese, and only slightly different from the grey goose species.

SIZE

The average Nene gander weighs 2,165 g (4 lb 13 oz) and his mate 1,930 g (4 lb 4 oz). However, from measurements taken of the captive flock at Slimbridge in England, we know that weight there varies during the year (Figure 10). Both sexes are heaviest at the start of the wing moult in April and lightest at the end of that moult; the male has replaced his wing feathers entirely by June, and the female by the beginning of July. They also put on weight again in the autumn or fall, but have lost most of it by January which is the start of the

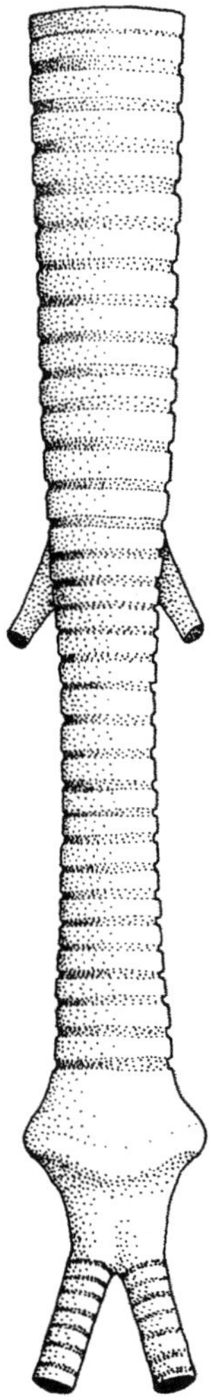

Fig. 9. Trachea, or windpipe, of adult Nene. Tim Halliday.

Slimbridge breeding season. This seems a rather unusual pattern, as most geese are heaviest as they prepare to breed. Weights of wild birds are not available for comparison, and the 52 captive Nene in Hawaii have only been weighed twice (10). These birds, like the

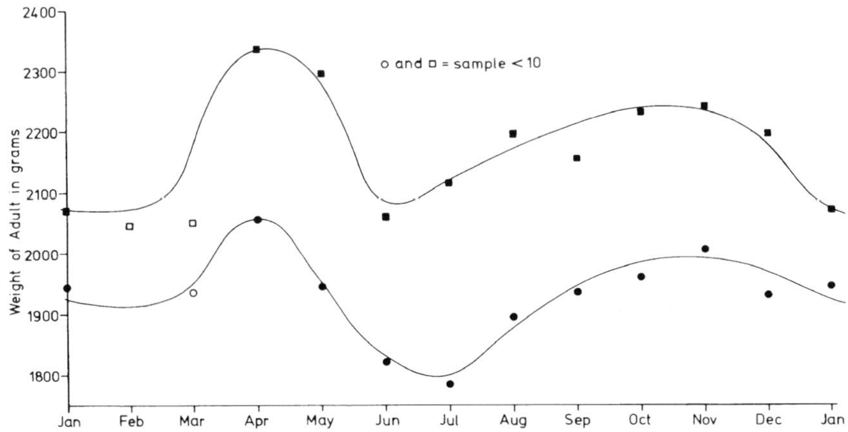

Fig. 10. Weight variation of adult Nene through the year at Slimbridge. The males are at the top. Both sexes are, on average, heaviest in April and lightest in June and July at the end of the wing moult. See Appendix 2 for details.

English ones, were lighter after moulting (which occurs about one month earlier in Hawaii). The question of weight is examined further in Chapter 4, and detailed statistics are given in Appendix 2.

STRUCTURAL ADAPTATIONS

The Nene is taller and more upright than other geese of similar weight, and has a slightly longer, down-turned bill (Figures 4 and 11), longer legs and shorter wings (Table 1).

The striking differences between the Nene and other geese are due to the species' modification for an island habitat. Its continental relatives spend much of their time grazing and swimming. There is little standing water or natural grassland in Hawaii, and specialisation in the Nene (which has, as already pointed out, a smaller range

TABLE 1 *Average measurements of adult Nene at Slimbridge (see Appendices 2 and 3 for details)*

	Males	*Females*
Weight	2165 g (4 lb 13 oz)	1930 g (4 lb 4 oz)
Head (bill tip to back of head)	94 mm (3·7 in)	89 mm (3·5 in)
Bill (tip to hairline)	39 mm (1·5 in)	37 mm (1·5 in)
Leg length	85 mm (3·4 in)	78 mm (3·1 in)
Wing length	378 mm (14·9 in)	361 mm (14·2 in)

than any other goose) has been towards a non-aquatic habit and to a feeding technique of standing upright while browsing on herbs and shrubs rather than reaching down to graze at the level of the feet. With this new feeding habit has come an increase in walking and climbing ability. A. H. Miller (96) was able to examine four specimens that had died at Keaau (Plate 1), in the collection of Mr Herbert Shipman, whose interest in the Nene, as we shall see later, was a crucial factor in their preservation. Miller drew attention to a number of adaptations that equip the bird for running and climbing over rough ground. It has a powerful leg action, with little 'waddle'

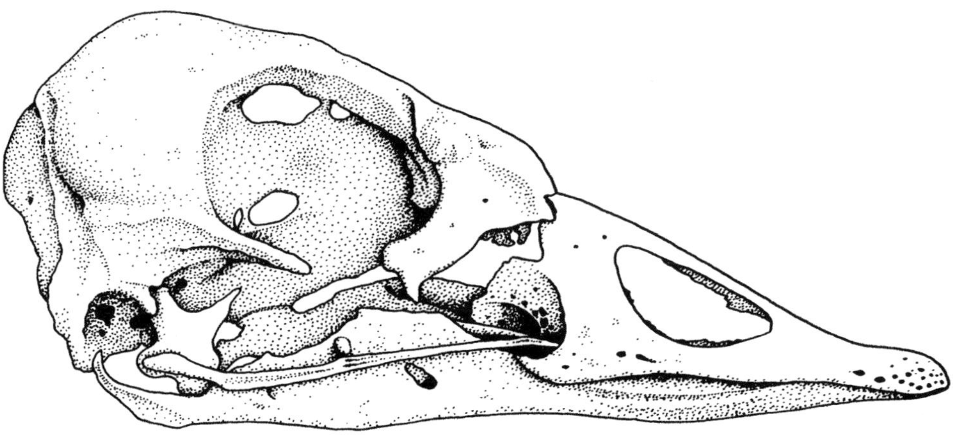

Fig. 11. *Skull of adult Nene.* Tim Halliday.

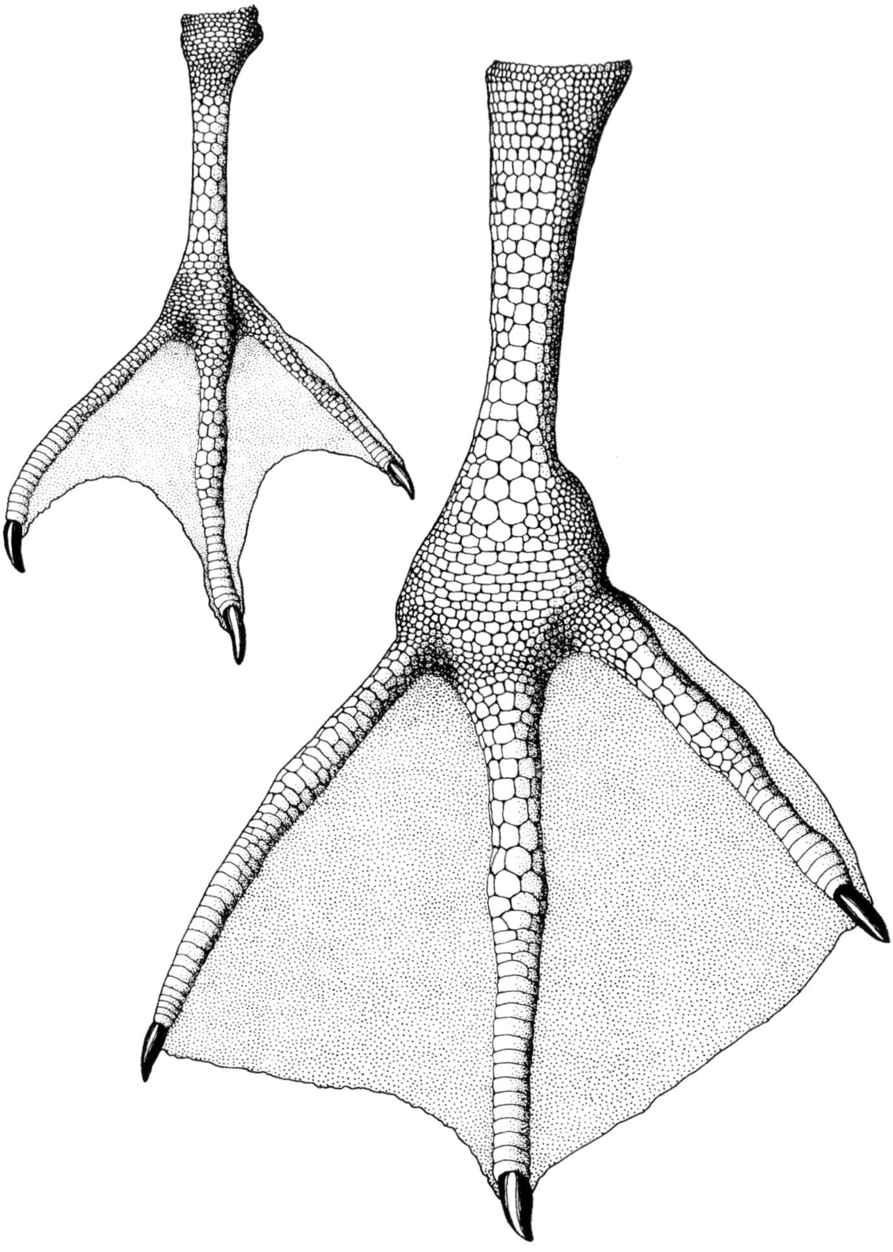

Fig. 12. The foot of the Nene (left) and that of the Atlantic Canada Goose. Tim Halliday.

or lateral motion of the limbs away from the line of progress. The pattern of musculature is not very different from other geese, but the total bulk of the leg in relation to the body is 25% greater, and the tendons of the toes are stronger. The Nene's foot (Figure 12) has flexible elongated toes, reduced webbing, large nails, and protective pads on the soles. Because of these specialisations the bird has sacrificed some of its (now unimportant) ability to swim efficiently. It also shows a reduction of about 16% (in comparison with other black geese) in the bones and muscles of the wing, presumably because it is non-migratory. Many birds of remote islands have reduced powers of flight – it is probably a positive advantage not to fly much, since a frequent flier risks being lost at sea. Nevertheless, Baldwin (14) suggested that feeding Nene flew several square miles during the course of a day, and some ringed birds are known to have travelled 112 km (70 miles) across the sea from the island of Maui to the large island of Hawaii (53; 105).

HABITAT

Hawaiian Geese live much of the year in the vegetated areas (called kipukas) that occur among the more recent lava flows, and where the only available standing water is in small temporary rain pools. The kipukas vary in size from less than an acre to several thousand acres. Rainfall is the most important factor in their creation out of the decomposing lava, and is essential for the establishment of vegetation, which begins with lichens, and is followed by ferns, grasses, herbs, shrubs, and trees. Less important parts of the Nene's former range included the lowlands on the leeward or western side of the island of Hawaii, moist grassland and open forests (14). The geese always avoided the dense humid forest of lowland and upland regions. However, in the 18th century, the Nene had a much wider distribution below 2,750 m (9,000 feet), and was found even at sea level. It is known that in those days the wild population used to descend below 330 m (1,200 feet) in the breeding season to take advantage of the growth of freshly sprouting vegetation as food for the goslings. In summer, as the weather grew hot and the plants less succulent, they would return to the uplands accompanied by their flying young to feed on berries (62; 104).

Baldwin (13) estimated that by the 1940s their range had been reduced to only 3,110 sq km (1,200 sq miles) and that most of this was in the highland areas. Today, on the slopes of Mauna Loa, birds are seldom seen below 1,500 m (5,000 ft) or above 2,500 m (8,000 ft).

The lower levels have been taken over by ranches and cultivated lands, in which the geese probably compete rather poorly with domesticated stock and exotic game birds, and where the introduced predators are certainly commoner. There is also a possibility that avian pox (Plate 23), carried in the saliva of introduced blood-sucking mosquitoes, makes land below 300 m (1,000 ft) unavailable to the Nene (83). This reduction in breeding range into what was, and probably still is, sub-optimal habitat may be of critical importance in the survival of the species in the wild.

FOOD

Baldwin in 1947 gave a detailed account of the food of the Hawaiian Goose. He found, by examining hundreds of droppings, that it fed on 31 different plants growing in areas that received between 380 mm (15 inches) and 3,040 mm (120 inches) of rain annually. During the hot dry summers, green food is scarce on the leeward or western slopes at low levels, but is abundant at all seasons at higher altitudes – so the geese are compelled to move upwards.

Baldwin's findings have been confirmed by more recent observations (10; 129; 137), and both studies showed that there is a marked preference for the leaves, buds, flowers and seeds of the dandelion-like gosmore (also called long-rooted catsear) *Hypochaeris radicata* (Figure 14). Seeds and leaves of grasses such as *Deschampsia nubigena* (Figure 13) and *Digitaria pruriens* (crabgrass) are also taken commonly. Ohelo *Vaccinium reticulatum* (and *V. peleanum*) and kukaenene* *Coprosma ernodeoides* (Figures 15 and 19) are the most important berries in the Nenes' diet, and it is probably from such juicy fruit that much of their water intake comes. They are preferred to another berry, pukeawe *Styphelia tameiameiae* (Figure 16), although Baldwin suggested that its hard 'stones' could serve as grit, or grinding agents, in the birds' gizzards. Popolo *Solanum nodiflorum*, a favourite food in Baldwin's study, is today uncommon in Nene habitats, but the geese are now seen to take the seed heads of introduced mesquite grass, or Yorkshire fog *Holcus lanatus*, which has become a widely sown pasture plant. In captivity, Nene show a decided liking for the sow thistle (Figure 17) or pualele *Sonchus oleraceus*.

Nene feed most during the early morning and late afternoon, and

* The Polynesian word kukaenene means 'nene-dung'. The plant was presumably named by the Hawaiians because the blue-black berries stain Nene droppings purple.

Fig. 13. Deschampsia nubigena, *an endemic Hawaiian grass and important Nene food, leaves, stems and seeds being utilised.* Sue Monden.

usually rest in the shade through the mid-day hours. This pattern of feeding is typical of all geese although, in a colder climate, most northern species take only a short rest around noon.

At present, food resources at the higher altitudes where Nene are found, are not considered to be the factor limiting their numbers.

Fig. 14. Hypochaeris radicata or gosmore (*Compositae*). *Seeds, leaves, stems and flowers are eaten by Nene.* Sue Monden.

Introduced food competitors, such as goat, sheep and pig, are neither numerous there nor particularly inclined to take the birds' preferred food plants. It has even been suggested that the rooting habits of the pig may encourage the growth of gosmore and other succulents. Goats and especially pigs disturb nesting geese, so may

Fig. 15. Coprosma ernodeoides *or kukaenene (Rubiaceae). The berries are eaten by Nene and stain their droppings purple.* Sue Monden.

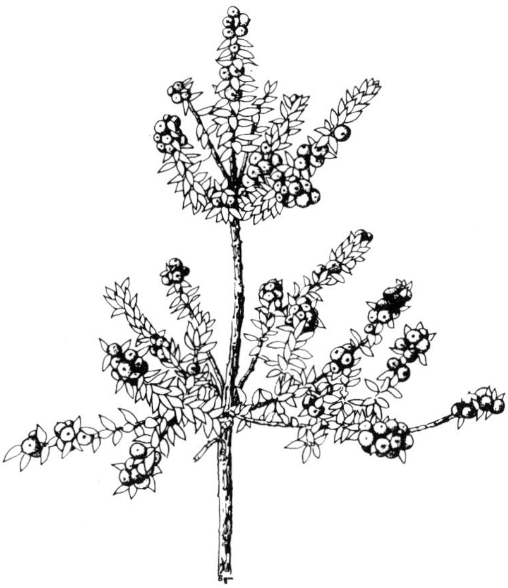

Fig. 16. Styphelia tameiameiae *or pukeawe (Epacridaceae). The fruits are eaten by Nene, the stones passing through the bird undigested.* Sue Monden.

still affect their numbers adversely although not, it is thought, by depriving them of their food directly. However, it is likely that, in high numbers, they attract feral dogs to the area and these mammals may be exceedingly damaging (137).

Fig. 17. Sonchus oleraceus, *sow thistle or pualele (Compositae). An important Nene food in the lowlands, and in captivity.* Sue Monden.

Fig. 18. Physalis peruviana, *cape gooseberry or poha (Solanaceae). Mentioned as a possible Nene food by Baldwin, who saw a foraging goose apparently feeding from the vines (14).* Sue Monden.

Fig. 19. Vaccinium peleanum *or ohela (Ericaceae). Abundant on Mauna Loa, where the berries are a likely Nene food (14).* Sue Monden.

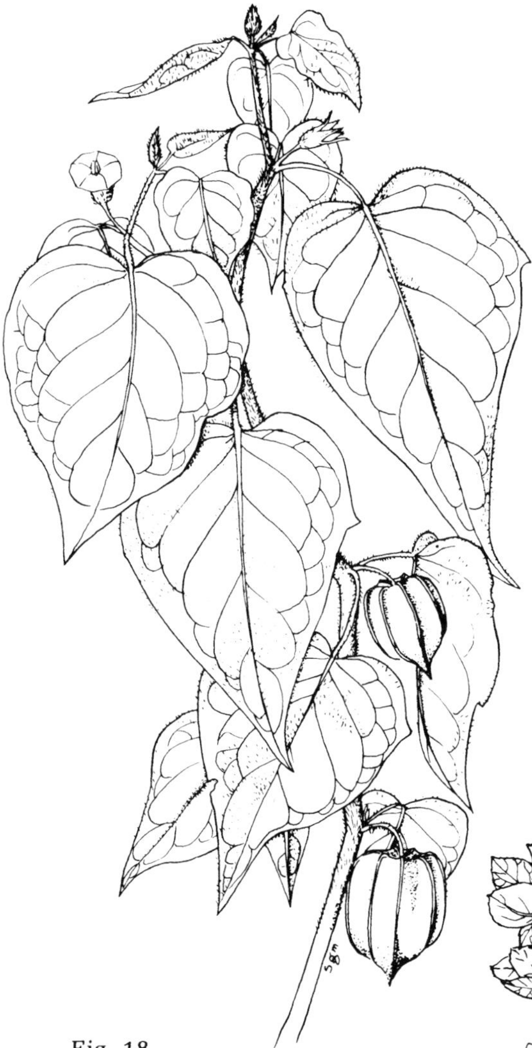

Fig. 18

Fig. 19

REPRODUCTION

The Hawaiian Goose differs from other black and grey geese in that it comes into breeding condition on a decreasing day length (21). Nesting in the highlands starts at the beginning of November, and active nests have been found as late as April. The factors controlling the length of this breeding season are considered more fully in Chapter 4.

Breeding territories are usually established in the vegetated kipukas because they provide the best nesting cover, as well as food. To some extent the use of these areas tends to isolate breeding pairs. Not much is known about how large the defended territory normally is. In captivity, the Nene is a fairly aggressive animal (aggressive enough to be described by Delacour (39) as ill-tempered) and is prone to demonstrate its rights over a breeding area obviously and vocally. We can assume from its captive behaviour that the use of a nesting territory in the wild, from which other Nene are excluded, is usual and a positive advantage to the survival of the pair.

(a) Display

A variety of aggressive displays are used by the Nene in defending itself, the most common being a bent-neck threat posture (Plate 7) which is accompanied by vibrations of the furrowed neck feathers (66). The goose faces an opponent with neck and bill pointing down at an angle of about 45°, and the quivering of the feathers becomes very obvious, not only because the plumage is arranged in vertical furrows, but because the buff coloured feathers are dark at their base. Vibrations also occur during low-intensity threat, when the head is held in a more relaxed and normal position (Plate 12). A particularly angry Nene uses a threat display in which the head is thrust, usually close to ground level, towards the enemy. This posture is much like the head-forward threat of the Canada Goose, and is almost always accompanied by calling with the mouth open and sometimes with the tongue raised (Plate 7). The pale neck plumage is not vibrated by an extremely aggressive goose, but some of the dark feathers on the back are raised, presumably to accentuate the birds' size.

Courtship, which results in the establishment of a pair-bond that tends to be life-long, takes place in the second year. As in all geese, it is an inconspicuous and lengthy process. The male walks rather stiffly ahead of the female, showing his white undertail (in other geese, this is done on the water). Once the female has accepted him,

1. *The small inbred flock of Hawaiian Geese held by Herbert Shipman at Keaau, Hawaii, prior to 1949. From this stock came the two pairs sent to Pohakuloa in 1949, and the three birds given to Slimbridge in 1950 and 1951. Photo: Herbert Shipman.*

2. *Male Hawaiian Goose Branta sandvicensis at Slimbridge. Photo: J. B. Blossom.*

3. *Heads of the four nearest relatives of the Hawaiian Goose. Top left: Redbreasted Goose Branta ruficollis. Top right: Black Brant B. bernicla orientalis. Bottom left: Lesser Canada Goose B. canadensis parvipes. Bottom right: Barnacle Goose B. leucopsis. Photos: J. B. Blossom; P. Wallis; G. Williams; J. B. Blossom.*

4. Top: A small kipuka surrounded by an aa lava flow. Keauhou Sanctuary. 9 January 1973. Photo A. J. Berger. Bottom: Nene sometimes are very inconspicuous on aa lava flows; note the pale neck and breast of the bird near the edge of the kipuka in the distance. This picture also shows the typical relationship between a lava flow and a kipuka. Photo: A. J. Berger 10 January 1973. Keauhou Sanctuary, Mauna Loa.

5. *A pair of Nene in the Keauhou Sanctuary, Mauna Loa, 9 January 1973. The coloured bands on the birds' legs indicate the year of release. Unfortunately, numbered metal bands were not used and, as the coloured rings were sometimes lost, there was confusion of information from re-sighted geese. Photo: A. J. Berger.*

6. *Various intensities of the 'upright threat' posture. Above left: low intensity display in male with ten-day-old gosling at Slimbridge. Above right: a higher intensity posture, with mouth opening and feathers at the nape being lifted. Bottom right: high intensity threat with open mouth, exposed tongue and ruffled nape feathers. Photos: J. B. Blossom.*

7. Intensities of the 'forward threat' posture. The birds at the top are the least likely to attack, while the gander at the bottom is the most aggressive. Photos: J. B. Blossom.

8. *A small part of the Nene rearing facilities at Pohakuloa, Hawaii, on the slopes of Mauna Kea; elevation about 2,000 m (6,500 feet). 11 January 1973. Photo: A. J. Berger.*

Moulting pair of Nene with young at Pohakuloa, 15 March 1950. The large water pools were not used in later years as the birds appeared not to require them. Photo: Hawaii Division of Fish & Game.

9. *Top: Nene pens at Pohakuloa, Hawaii, 15 March 1950. Photo: Hawaii Division of Fish & Game. Bottom: Ah Fat Lee with a pair of Nene and their four downy young; Pohakuloa, about 1955. Photo: Hawaii Division of Fish & Game.*

10. *Top: A gander standing guard while the goose is incubating on a nest nearby. Keauhou Sanctuary, 10 January 1973. Photo: A. J. Berger. Bottom: David H. Woodside of the Hawaii Division of Fish & Game looking at a Nene nest with two eggs in the Keauhou 2 Sanctuary; 11 January 1973. The goose flushed as our jeep drove by, but she had covered the two eggs with down. The clump in which this goose built her nest consisted of pukeawe, ulei, pilo, and aalii. The slope of Hualalai rises on the horizon. Photo: A. J. Berger.*

11. *Top: If an incubating goose has time, she covers the eggs with down feathers that line the nest before she leaves. In this instance, the goose was alerted to our approach by callnotes of the gander, who undoubtedly saw us before we saw him. The behaviour of both birds, however, made it relatively easy for us to find the nest, which contained two eggs. Keauhou Sanctuary, 9 January 1973. Photo: A. J. Berger. Bottom: A wild Nene nest with four eggs. Keauhou Sanctuary, 10 January 1973. Photo: A. J. Berger.*

12. 'Wigwam' constructed at Slimbridge as wind-proof nesting sites for Hawaiian Geese. The gander is on guard outside, and is showing a low-intensity threat posture during which the neck feathers are often vibrated. Photo: Philippa Scott.

. Above: Female Nene with week-old ·slings at Slimbridge. Photo: Philippa :ott. Right: Week-old Canada Goose gos- ıg for comparison. Photo: The Wildfowl ·ust.

14. *Above: Wild Hawaiian Goose with six-week-old young. Photo: David Woodside. Below: Five-week-old goslings at Slimbridge with their moulting parents. Photo: J. B. Blossom.*

15. *Adult and immature Nene ready for release into the wild. Pohakuloa, Hawaii, 16 March 1960. Photo: Hawaii Division of Fish & Game.*

16. *Flock of Nene at Slimbridge. Photo: The Wildfowl Trust.*

17. *Crating the first shipment of Slimbridge-bred Hawaiian Geese destined for release on Maui. Photo: J. V. Beer.*

18. *Above:* Typical habitat near the release pen in Keauhou 2 Sanctuary on Mt. Hualalai. Mauna Loa, elevation 4,169 m (13,677 feet), lies in the background; the summit has only a thin snow cover. 11 January 1973. Photo: A. J. Berger. *Bottom:* Because of the abundance of the introduced mullein in the Keauhou 2 Sanctuary, it often is difficult to locate Nene; there is one Nene in this picture. Mauna Kea, elevation 4,205 m (13,796 feet), lies in the background. 11 January 1973. Photo: A. J. Berger.

19. *Top: David H. Woodside and Ernest Kosaka of the Hawaii Division of Fish & Game stand near the release pen in the Keauhou 2 Sanctuary on Mt. Hualalai. Two Nene are still in the pen; 45 had been released here during September 1972, but many had to be returned to Pohakuloa because of damaged wing feathers. 12 January 1973. Photo: A. J. Berger. Bottom: The release pen in the Keauhou Sanctuary on Mauna Loa. 9 January 1973. Photo: A. J. Berger.*

20. A female Nene at Slimbridge in full wing moult. Note the abdominal swelling that appears apparently quite normally at this time. Photo: E. E. Jackson.

21. The preen gland is situated just above the tail. The bird in the upper photograph is nibbling the gland to stimulate oil flow. Photo: P. Wallis. In the lower picture, a moulting Nene is preening its growing wing feathers. Photo: J. B. Blossom.

22. *The recessive genetic character known as 'hairy' or 'cottony' down affects a high proportion of Nene goslings. Here, the normal goslings (on the left in the top photograph, and on the right in the lower picture) can be seen to have much denser, more fluffy down. Slimbridge, 18 March 1969. Photos: The Wildfowl Trust.*

23. *Top left and right; Lesions of possible avian pox on the sole of the foot of a Nene at Slimbridge (83). Below are pictures of a normal foot for comparison. Photos: M J. Brown & G. Williams.*

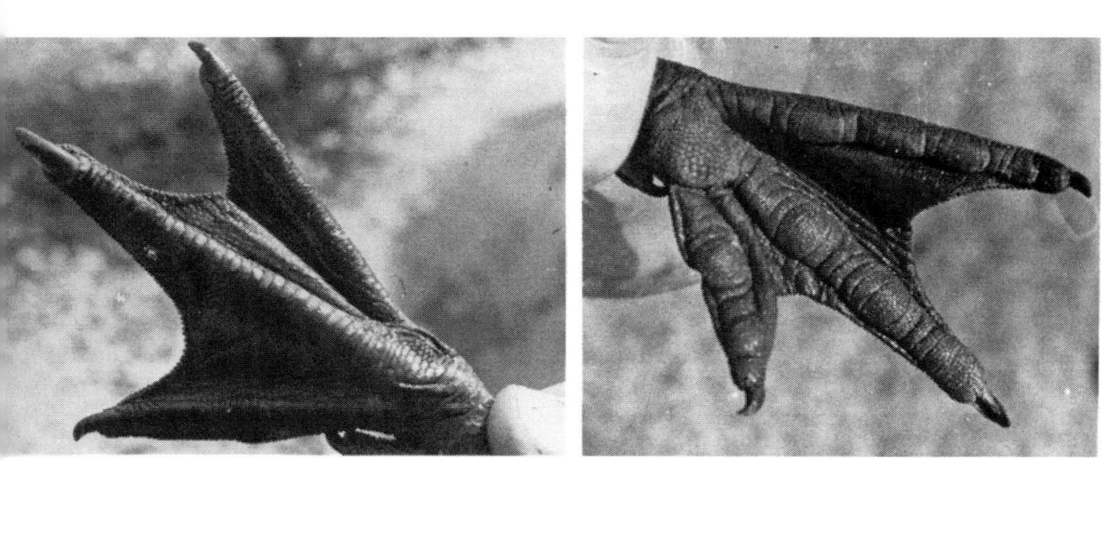

24. Nene dibbling in water and showing the deeply furrowed neck feathers. Photo: J B. Blossom.

they will engage in mutual 'triumph ceremonies'. This display is an important part of courtship; in it the gander, after threatening or attacking an opponent, runs back to his potential mate and calls loudly. He and she call or cackle almost directly into one another's ears, making intermittent sideways and vertical head movements more or less directed towards the opponent.

Unlike other true geese, the Nene copulates on land or, in captivity, on the shoreline. Johnsgard (66) noted that young captive geese usually attempt copulation at the water's edge, and once he observed two young birds trying to mate whilst swimming. In all cases, the precopulatory display consists of head-dipping movements, just as in other geese, and these are clearly derived from actions used in bathing. Each bird extends its neck and head along the ground, then rapidly pulls back, at the same time turning slightly, just as a bathing bird would. The two birds tend to synchronise their actions more and more closely, and finally the female goes prone and the male mounts her. When copulation is completed, the male opens his wings and pulls back the female's head; both birds then call and drop their wings slightly while extending their heads and necks almost vertically. The Nene's habit of mating on land is another example of its adaptation to an environment in which water is scarce. The bathing movements used in display, however, show that the species must have evolved from an ancestor that in the distant past copulated while swimming.

(b) Nest sites

The first wild nest ever seen by biologists was discovered on 9 November 1956 by Dr William H. Elder and David H. Woodside on the ranch known as Keauhou on the slopes of Mauna Loa (49). Since then, many nests have been found (Plate 10), mostly under pukeawe bushes and other shrubs, where the bird scoops out a shallow depression in the litter.

Only the female builds the nest, although one or two captive males at Slimbridge have been seen helping to add material. The gander guards the site from an elevated lava outcrop, or similar feature, some distance away (Plate 10). From there, he has an unobstructed view of his mate on the nest, as well as a commanding view of the surrounding countryside. When danger approaches, he gives a soft call. In the early stages of incubation, the goose will then leave the nest by walking low over the ground and not stand upright until she is some distance from it. This ruse is successful in concealing the site, since a predator usually searches the area where the goose first appears. During later stages of incubation, the bird will remain motionless on her clutch, and is remarkably inconspicuous. Unless a thorough search is made, a human is apt to walk right past. The easiest way to find a nest is to go to the gander's station, which can be readily identified by the piles of droppings, and to search from there.

There is evidence that breeding pairs return to the same kipuka every year (49). If the eggs are hatched successfully, but not otherwise, the same nest site may be used for several seasons in succession. Henshaw (62) observed in 1902 that the fact that Nene return to the same site was well known to the natives who, 'when once they find a nest, never fail to return the following year to secure the young.' Wild pairs seem to remain together until a partner is lost through death, and certainly captive pairs and trios (one male and two females) are unlikely to part once they have bred.

(c) The egg

The egg, at about 150 g (and with a range of 120–180 g), is large in relation to the weight of the female bird. As Lack (87) pointed out, this is normal in island forms in comparison with their continental ancestors. Three hundred eggs measured at Slimbridge were on average 83 mm long by 56 mm wide, which is somewhat larger than the 80 × 50 mm given by Delacour (39). It is possible that Delacour's eggs came from birds that had been many generations in captivity and were producing smaller eggs than they do in Hawaii; indeed, it is likely that Hawaiian eggs are always larger than those laid in Europe (see Appendix 4), a point that is further discussed in Chapter 4. The first published account of three wild eggs was written by Henshaw (62) in 1902, who described them as creamy white and measuring 3·37 × 2·42; 3·32 × 2·45; 3·40 × 2·18 inches (an average of 85·4 × 59·7 mm). Bryan (36) of the Bishop Museum, Honolulu, reported in 1906 that they were oval, of good firm texture, pure white in colour, with a smooth even surface, and measured 3·30 × 2·45 inches (84 × 62 mm).

The shell weighs about 14 g or 10% of the fresh egg, and the yolk makes up another 41% (88) – a normal proportion in waterfowl.

(d) Clutch size

The clutch, according to Bryan (36), consists of two or three eggs, and rarely five. Henshaw (62) said that the bird produced '3–6 eggs, the former being doubtless the more usual number.' Delacour (39), on the other hand, gave the clutch size as five to eight, which is

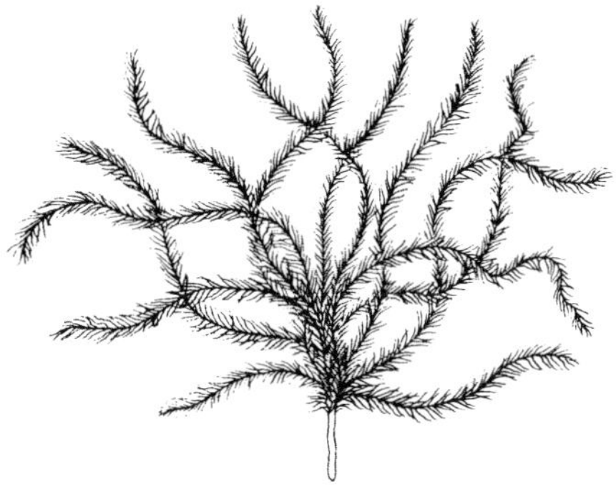

Fig. 20. A down plumule from the breast of the female Nene. Tim Halliday.

certainly too large. The clutch is, in fact, small for a goose – an additional feature shown by many island waterfowl (87). Clutches at Pohakuloa and Slimbridge normally contain four or five eggs, and vary from three to six; at times birds lay only one or two eggs, but these are seldom incubated.

(e) Incubation

Incubation lasts 30 days (rarely 29 or 31 days); longer than in any other true goose. Only the female incubates, starting almost as soon as the last egg is laid, thus ensuring that the whole clutch hatches simultaneously. She pulls down (Figure 20) from her breast and with this (plus some vegetation) covers the eggs when she leaves on foraging trips, probably once or twice a day (Plate 11). A goose that is, or has recently been, incubating can be recognised by the almost featherless brood-patch on the abdomen where the eggs are brought into contact with the warm skin. The 1834 observations of Lord Stanley (121) on a captive bird describe typical Nene behaviour: 'During the whole of incubation there could not be a more attentive nurse, and indeed she could not well help it, for the male, if she seems inclined to stay out longer than he thought right, appeared, by his motions, to be bent on driving her back, nor was he satisfied till he had accomplished his object'.

If the clutch is infertile or dead, the female will sit for considerably longer than a month. If her eggs are lost, or the nest deserted, a Nene in captivity may lay again. The time between the loss of eggs and re-laying varies with the length of time that the bird has been sitting:

if the eggs disappear as soon as they are laid, the female may nest again in three weeks; if they are lost close to hatching, it may be two months before she lays again. There is little evidence to indicate whether wild Nene do re-nest. A build-up of food reserves would be needed in order to produce further eggs, and a shortage of high-protein and high-calcium food might make a second clutch unusual in the wild. However, in captivity at Pohakuloa and at Slimbridge, captive Nene often re-nest.

(f) The gosling

The hatching of the entire clutch takes about 24 hours. The goslings then weigh on average 100 g (90–115 g) (see Appendix 5 for details). Weight loss during the 30 days incubation is therefore of the order of 35 g, or a quarter of the original weight (excluding the shell). This is accounted for by the membranes left in the shell after hatching, by water loss and by general metabolism.

As a comparison, it is interesting that 24 newly hatched Atlantic Canada Geese at Slimbridge weighed only 118 g on average with a range of 92–143 g. Thus there is considerable overlap in the gosling size of these two related species although, when adult, the Canadas weigh well over twice as much as the Nene. The heavy gosling is related, of course, to the heavy egg. The evolutionary pressure behind the production of large offspring is not understood, but Lack (87) supposed that island conditions might be unusually rigorous and that a bird that is large at hatching is at a definite advantage in conditions where food is not abundant or readily available.

The gosling is dark grey-brown, quite different in colour from the yellow Canada gosling (Plate 13). The head and neck are patterned: the crown and nape dark, with an irregular dark stripe running through the eye, and a dark spot on the 'ear'. The chin and throat are lighter, and a pale band extends across the forehead to the eyebrows. The bill, legs and feet are black.

Young Nene are charming creatures, fluffy, bright-eyed and even less fearful of their surroundings than other geese. Normally, the gosling has a covering of dense close-packed down but, in captivity at Pohakuloa and Slimbridge, a few have hatched with down that is thin. These goslings are known in England as 'cottony' and in Hawaii as 'hairy'. The condition seems to result from a recessive gene that causes some of the short down plumules to be missing so that the longer ones appear more prominent (Plate 22). The young look scruffy, and probably are not waterproof. In captivity, with adequate heat, they are at no particular disadvantage, and in full

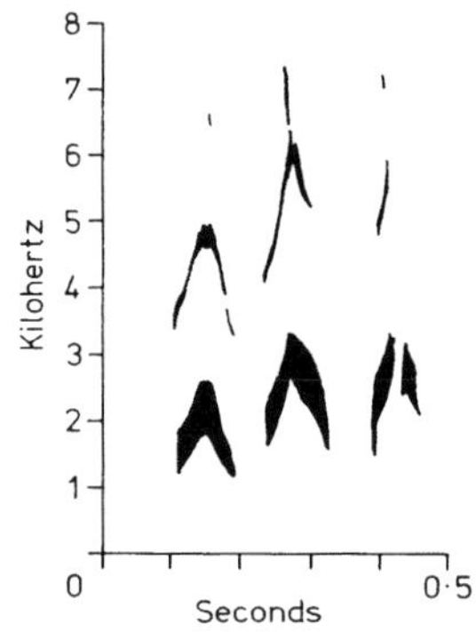

Fig. 21. *Tracing of sonagram of contact call of Nene gosling.*

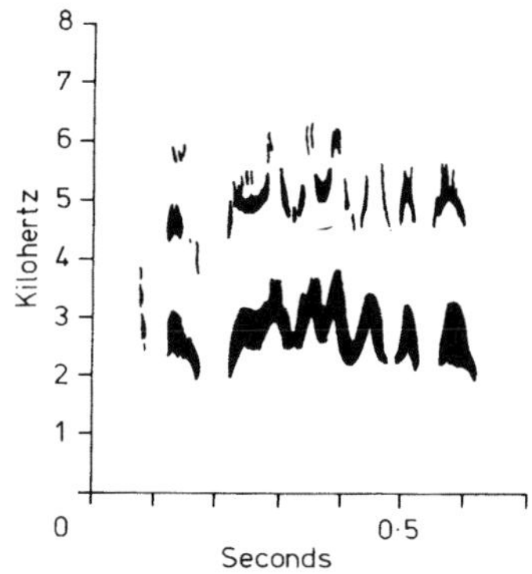

Fig. 22. *Tracing of sonagram of sleepy call of Nene gosling.*

plumage are indistinguishable from normal birds. The trait is considered further in the final chapter.

The voice of the gosling is a most important part of its behavioural equipment (76). With it the young bird can indicate, among other things, that it is lost, sleepy, or content, and thereby elicit appropriate actions from its parents. 'Pleasure calls' are the soft sounds heard when the family feeds, preens or moves together, and they keep the birds in contact with one another. The notes are grouped in twos and threes; Figure 21 shows what they look like on a sonagram. 'Greeting' is similar, but louder, and the notes have more of a rising and falling cadence. The young bird greeting its parent stretches its patterned head and neck forwards towards the older bird as it calls. When feelings are intense (when, for instance, it suddenly finds its parents after having been lost) it opens its mouth, raises its tongue and repeats the syllables in an almost ecstatic manner. 'Sleepy calls' are given when the goslings are tired, and they probably encourage the brood to synchronise their rest periods. Six contact notes are joined together in an ascending and descending trill (Figure 22) of which the fundamental frequency, like that of the 'pleasure call', is mainly low. 'Distress calls', on the other hand, are loud and high. They are given when the youngster is lost, cold or hungry. A deserted gosling stands erect, with neck stretched upwards, mouth open, and with its down on end, giving one call to every breath (the notes are therefore evenly spaced – see Figure 23) and making itself as conspicuous and noisy as possible. The sounds are easy to locate, and parents will go in search of their crying goslings.

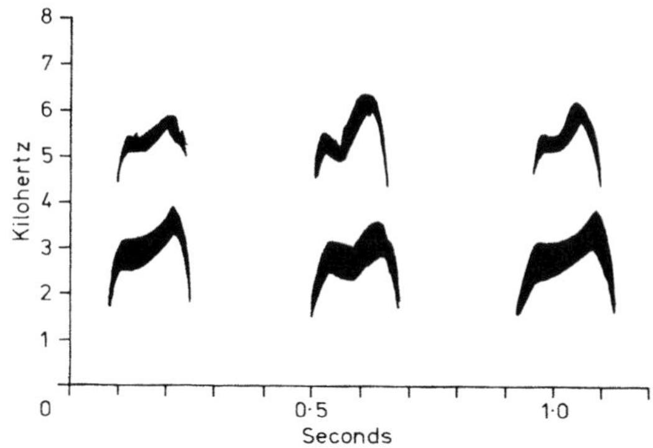

Fig. 23. Tracing of sonagram of distress call of Nene gosling (from Kear (76)).

There is probably high mortality at the juvenile stage in the wild. Even in captivity, something like 20% of goslings die before they fledge (72), and in the Hawaiian habitat, with its dangerous lava slopes and occasional, heavy rainfall, young must be lost fairly frequently. Both parents guard the young, leading them to suitable feeding grounds, and remaining with them until the next breeding season. The fledging period, like incubation, is longer than that of any other goose (Lack has pointed out that there is a general association between these two periods in birds). In Hawaii, Nene first fly when they are about 10–12 weeks old (47). Before they do so, the adults moult their feathers, the male slightly before the female (see Figure 10). During the moult, which may occur in the same kipuka where the babies hatched, the birds lose all their wing feathers at the same time, and are flightless for four to six weeks (47). This moult, like the breeding cycle, is probably controlled by daylength, although the timing is variable and depends upon the condition of the bird. Geese in their second spring, and failed breeders, may moult earlier than parent birds. While they are flightless, geese are especially vulnerable to predators, and this may have been the stage at which the introduced mammals did their most serious damage to wild Hawaiian stocks. Unlike other geese, the Nene cannot resort to lakes and rivers for protection, as there are none. Yealland (139) noted that in captivity in Hawaii the female moulted when the goslings were 16 days old, and she thus lost some of her capacity for brooding.

Even when the young are fully feathered and flying, they are

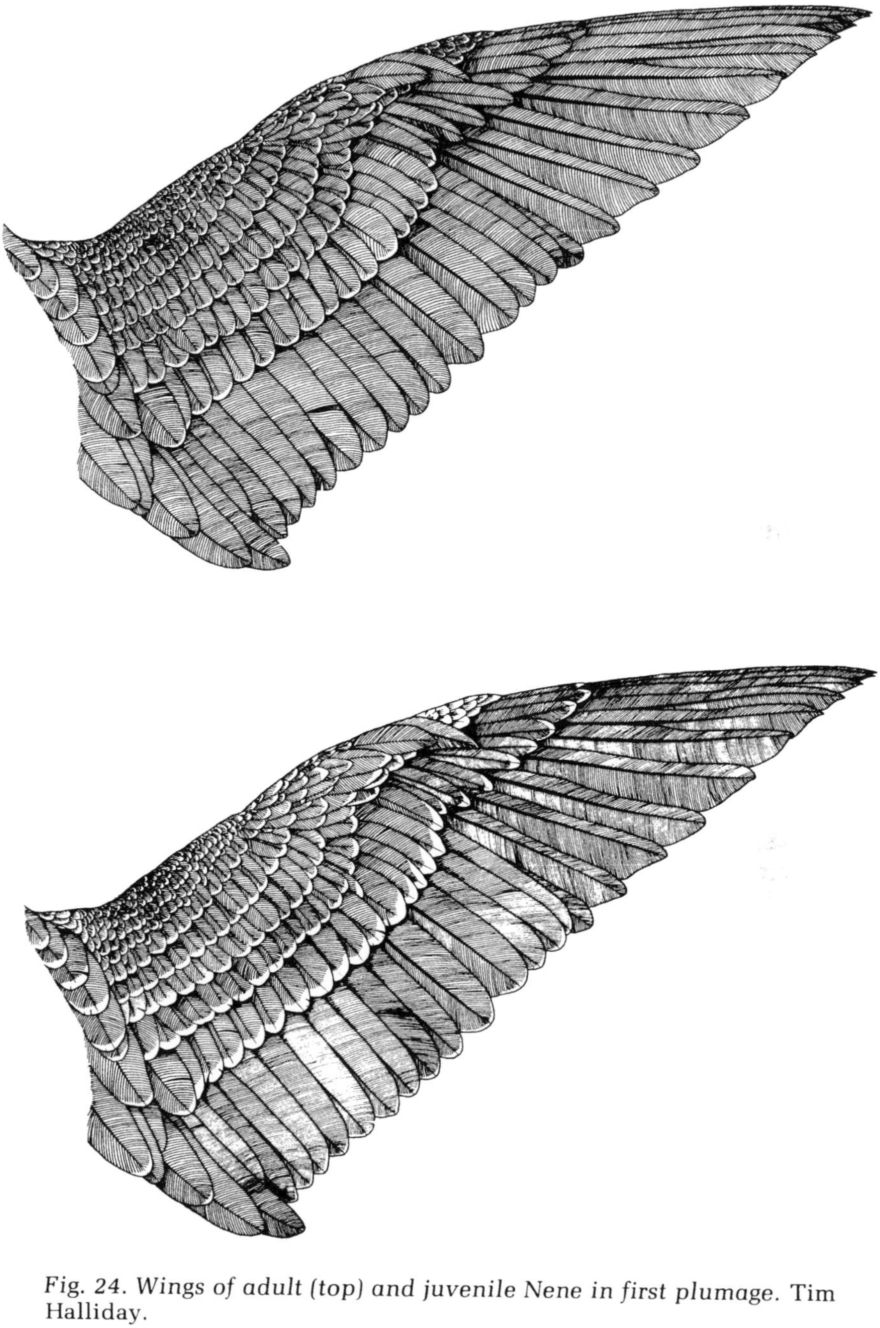

Fig. 24. Wings of adult (top) and juvenile Nene in first plumage. Tim Halliday.

easily distinguishable from their parents. The voice is still high and squeaky, and the plumage is not so bright as that of the adults, the general tone being more grey than buff, with all the black and brown markings less clearly defined (Figure 24). The furrows of the neck feathers are already visible in the young bird, but the yellow-buff areas are a greyish black. Miller (96) drew attention to the fact that the juvenile Nene resembles the Canada Goose (the supposed ancestral stock) more than does the adult (Figure 4).* During about six weeks of their first summer, the young birds moult their juvenile feathers, and acquire adult plumage and an adult voice. The wing and tail feathers remain until the following year.

The food of the young gosling has never been investigated, but is almost certainly vegetarian, and similar to that of the adult.

THE FLOCKING PERIOD

From June to early September, wild Hawaiian Geese flock in pairs and families, and move daily from a roost to a feeding ground, just as northern wintering geese do. Their behaviour at this time has been examined by Woodside (135) and Elder (47). On 21 July 1955, John Ah San, a State forest ranger, saw a flock of seven wild geese in the Kipuka Ainahou area. David Woodside, a game biologist with the Division of Fish and Game, studied the habits of these Nene for a period of ten weeks. He discovered that the flock contained 28 birds, and that they spent the daytime hours feeding and loafing on the 1881 and 1852 lava flows located just to the north of Keauhou ranch. This was later found to be one of the primary nesting areas on the eastern slopes of Mauna Loa adjacent to Volcanoes National Park. The 1852 lava flow, where the birds spent most of their time, was formed of *aa* lava (very rough angular blocks), which was covered with moss, small shrubs, and scattered ohia trees *Metrosideros collina* 3–7·5 m (10–25 feet) high. Ohelo and kukaenene berries, favourite foods of the Nene, were also abundant, but grasses were practically non-existent.

Between 16.30 and 18.30 hours, the Nene left the feeding ground and, in groups of two to a dozen, flew north across the Saddle Road between Mauna Loa and Mauna Kea to the Puu Oo ranch (owned, coincidentally, by the same Herbert Shipman who had captive Nene

* In 1954, one Slimbridge gosling developed a completely black neck rather like a Canada Goose (7). This gander survived for only six months, so it was not possible to determine the genetic nature of the variation.

at Keaau). Here, halfway up the eastern slope of Mauna Kea, the birds grazed rattail grass and roosted for the night. In the morning, between 07.30 and 09.30, the geese left in small groups for their daytime feeding grounds.

While Nene are in summer flocks, the young birds select their future mates. Maturity in the wild was thought by Elder (47) to be reached at three years, but in captivity, at least, some geese breed when they are only two.

PREDATORS

The cryptic plumage of the Nene, and its behaviour on the nest, seem to suggest that it had some natural Hawaiian predators before the introduction of mammals – and predators, moreover, that hunted by sight. When the goose stands still, it is extremely hard to pick out against the lava rocks (Plate 18). Perhaps the Hawaiian Hawk or Io *Buteo solitarius* and the ground-nesting Short-eared Owl or Pueo *Asio flammeus sandwichensis*, now uncommon, were natural enemies of young goslings, or even adults. However, Elder (47) commented that 'when pueo and io were once abundant, so were Ne-Ne; so the native Hawaiian predators . . . cannot be implicated in any way in the decline of the goose.' When approached to within about 180 mm (200 yds) by a human, Nene start calling and often move to an elevated position, thereby making themselves particularly easy to shoot (135). Dogs sometimes catch healthy, full-winged wild birds – and not only those sitting on nests – which perhaps suggests that Nene tend to 'freeze', instead of flying from a source of danger. In captivity, they are unusually attracted to dogs, seeming to have little residual instinctive fear of them.

P.S. 1950.

3: The Nene in Captivity (especially in Hawaii)

On 11 June 1833, an Hawaiian Goose was exhibited at a meeting of the Zoological Society of London. It was one of a pair living at the Society's Gardens, to which they had been presented by Lady Glengall in 1832 (3; 115). Mr N. A. Vigors, Secretary of the Zoological Society, characterized it as 'a species of *Barnacle Goose*, by the name of *Bernicla Sandvicensis*, and pointed out its distinguishing marks. He also observed on the general resemblance in the distribution of colouring which occurs in the species of *Bernicla** and in those of many other groups of *Birds*' (3).

Lord Stanley, the 13th Earl of Derby, sent a young Nene, hatched at Knowsley Hall, near Liverpool in Lancashire, to be shown at a meeting of the same Zoological Society on 27 May 1834, and he described his experiences with the geese (121):

'Through the kindness of John Reeves, Esq.,† I received at Knowsley a pair of these birds on the 15th of February, 1834. They did not at first, when turned out on the pond among the other waterfowl, appear to take much notice of each other; but some workmen being at the time employed about the pond, one of the birds (I think, from recollection, it was the male), seemed to have formed some sort of

* At the 1833 AGM of the Zoological Society of London held on 29 April, Council's Report (127) of new aquisitions lists the Sandwich Goose as *Anser Sandvicensis*. Presumably, Vigors later changed his mind and decided that the bird was a black goose rather than a grey *Anser* species. *Bernicla*, the genus to which he assigned the Nene, is the latinised form of the English word 'barnacle' (before geese were known to breed in the arctic, they were supposed to hatch from shellfish of the same name). Later, a taxonomist at the British Museum, Count Salvadori, noting how different the Nene was from other geese, changed its scientific name to *Nesochen*, Greek for 'island goose' (112). This removed the bird from the group to which its undoubted relatives belonged, and it was restored by Delacour and Mayr in 1945 (40). Meanwhile, the name for black geese had also been changed, from *Bernicla* to *Branta*, a word apparently deriving from the Greek *brenthos* meaning mysterious waterbird!

† We have been unable to establish whether this is the naturalist John Reeves, FRS, FLS (1774–1856), who spent many years in China and collected natural history specimens for British museums and gardens, including Knowsley Hall. He returned to England in 1831, bringing with him the first live Reeves's Pheasant *Syrmaticus reevesi* and settled at Clapham.

attachment to one of the men working. Whenever he was present the goose was always near to him, and whenever absent at his dinner, or when otherwise employed, the bird appeared restless, and gave vent to its solicitude by frequent cries, which as well as the anxiety, always ceased with the reappearance of the workman.

'The man having frequently occasion to pass through a door, which was obliged to be kept open, it was feared that the attachment of the animal might lead to its following its friend, and that on its exit, it might fall in with and be worried or stolen by vermin, and in consequence the pair of geese were confined in one of the divisions adjacent to, but divided from, the pond, on February 26th.

'Within this small enclosure, in the sheltered half of it, in one corner, stood a small hutch, in which the female on the 5th of March laid her first egg. Till within a few days of that period no alteration took place in their manners, but it then became obvious that the male was jealous of intruders, and would run at and seize them by the trowsers, giving pretty sharp blows with his wings; but this always ceased if he observed that the female was at some distance, when he would instantly rejoin her: his return to the female was always accompanied by great hurry and clamour, and much gesticulation up and down of his head, but not of the wings. Three other eggs followed on the 7th, 9th and 11th of March. The eggs were white, and very large in proportion to the size of the bird, being, I should imagine (for, having no proper scales at hand, I did not weigh or subtract any of them, hoping that more might be laid), fully equal to those of the *Swan Goose* or *Anas cygnoides*. The goose also surprised us by the rapidity of her operations, for we were hardly aware of the fourth egg having been laid that morning, when it was evident that she had begun to sit . . .

'On the 12th April the eggs began to chip, and on the 13th two goslings were excluded; but it was found that the mother had pushed from under her the other two eggs, which were consequently taken away and put under a hen, though, as one was very nearly cold, little hopes of any success with that were entertained, and it was in fact never hatched, but probably died in consequence of the removal by the goose at an important moment. On the morning of the 14th it was ascertained that she or the male, who always now sat close beside her in the box, had killed one of the two she had at first hatched, for it was found dead and perfectly flat.* The fourth egg, which was put

* The skin of this gosling, the first ever hatched in captivity, passed to the collections of what is now the Merseyside County Museum, Liverpool, on the death of Lord Derby and is there today.

under the hen, was assisted out of the shell, and appeared weakly from the first, and as its mother had lost one, we put it to her, in hopes it would do better than with its nurse. She took to it at first very well; but subsequently, both the parents beating it, it was returned to, and well cared for, apparently, by its nurse, but died on the 20th, having received some injury in one eye, either from the old ones, or perhaps from the hen scratching, and thereby hitting it. The remaining gosling is doing very well, and appears strong and lively, and the parents are extremely attentive to it; and I have little doubt that these birds may easily be established (with a little care and attention), and form an interesting addition to the stock of British domesticated fowls.

'In its general appearance, and its Quaker-like simplicity of plumage, it seems to approximate most to the family of the *Bernacles*; but it appears to have almost as little (if as much) partiality for the water as the *Cereopsis*.'

The Nene did become successfully established after this first captive breeding.* The pair at the London Zoo bred the following year, and Sclater (115) recorded 38 clutches laid there between 1835 and 1871, the earliest hatching on 20 March and the latest on 30 May. Birds were widely distributed to collections in England and on the Continent (38; 39). Hutton (65) in 1871 even listed them as among birds bred in New Zealand for release into the wild, 'but not yet turned out.' Although Nene were raised by a number of people, few aviculturists published their experiences. F. E. Blaauw (22) of the Netherlands, one of those who did, wrote that 'the next birds to lay eggs in 1903 were a pair of Sandwich-Island Geese (*Neochen* [*sic*] *sandvicensis*), a species which has become extremely rare in Europe of late years. These birds are kept at Gooilust in a small grass-grown enclosure, with plenty of shrubs and a wooden shed in it. As they are not happy in frost and snow, I have them shut up in the shed every night after winter has set in. One good result of this arrangement is that the birds usually build a nest in the hay which covers the floor, and this makes it possible to protect them and their eggs from the cold weather that often prevails at the early season when they are accustomed to lay.

'Last February five eggs were laid. These were all hatched, and

*When the 13th Earl of Derby died in 1851, his unique collection of 1,272 birds and 345 live mammals was dispersed. The auctioneer's catalogue (4) included 70 captive-bred Passenger Pigeons *Ectopistes migratorius*, four Carolina Parrakeets *Conuropsis carolinensis* and five 'Sandwich Island Geese'. The pigeon and parrakeet were extinct by 1914; the goose was saved, but only just in time.

during the whole time of incubation the male was constantly on the watch beside the female, running with great fury at everyone who came near.'

Blaauw was the last breeder to rear the Nene successfully before the 1950s (38). The cause for its decline is apparently not known, but male infertility due to inbreeding is likely to have been important. Dr Jean Delacour had the last European specimen at his collection at Clères, France. This gander had been hatched at Gooilust, and was the only Nene left upon Blaauw's death in 1936. It vanished from Clères at the time of the German invasion in 1940 at the reported age of 42 years (39). So ended over a hundred years' experience of Nene propagation in Europe.

EARLY PROPAGATION EFFORTS IN HAWAII

At no time during the past 150 years have the government and business leaders of Hawaii shown any real appreciation of, or effective concern for, the native flora and fauna. If they had, there would not now be so many extinct and endangered species of birds, snails, and plants, or so few native forests.

The lack of concern shown since the 1890s has not been from ignorance of the facts. Admittedly the first important list of Hawaiian birds, compiled by Sanford B. Dole in 1869 (41), contained a number of errors and included only about half of Hawaii's unique land birds. Dole himself (42) published a revised and improved list in 1879.

It seems likely that a renewed interest in Hawaiian birds was kindled by Charles Darwin's writings on evolution. Professor Alfred Newton of Cambridge University was responsible for the visit to Hawaii of Scott B. Wilson in April 1887. When he returned to England more than a year later, he took with him a large collection of bird skins. This collection formed the basis for *Aves Hawaiiensis*, which was published in sections between 1890 and 1899 (133). Wilson returned to Hawaii in 1896, and preserved entire birds in spirits. These specimens were studied by the famous German anatomist Hans Gadow, then resident in England.

Lord Walter Rothschild also became interested in Hawaii because of his association with Professor Newton. Rothschild, a member of the famous banking family, sent his bird collector Henry Palmer to Hawaii in 1891; there Palmer hired a young New Zealander, George C. Munro, as an assistant. They first made an expedition to the Leeward Islands, followed by extensive collection on the main islands, where they discovered a number of new species. The collection made by Palmer and Munro served as the basis for the elaborate *The Avifauna of Laysan and the Neighbouring Islands*, published by Rothschild in three parts between 1893 and 1900 (111).

We owe to R. C. L. Perkins, however, virtually everything that we know about the general habits and feeding preferences of Hawaii's birds during the 1890s. Perkins, a trained entomologist and fine field biologist, was sent to Hawaii by joint sponsorship of the Royal Society and the British Association for the Advancement of Science. Perkins reached Hawaii in 1892, and continued his field studies for ten years. His contribution in the volume on birds in *Fauna Hawaiiensis* was published in 1903 (104).

All of these authors mentioned introduced predators, the destruction of the forests by introduced grazing mammals, and the resultant extinction of Hawaiian birds. Wilson made a special appeal in 1890 to the land-owners and to the Legislature of Hawaii to unite in protecting their country's birds. 'I would suggest that not only should forest-lands be fenced in so far as practicable, but that *no* exotic birds should be introduced. Several species of Hawaiian birds, which were to be found in Cook's time, and others which were obtained even so late as 1840, have become extinct, and it would not be rash to say that ere another century has elapsed but few native species will remain.'

Henshaw (62) stated that he had prepared his list of Hawaiian birds 'chiefly with the hope that the meagerness of our knowledge respecting this subject may thereby be made more apparent, and thus that island observers may be stimulated to enter this very interesting and fruitful field – a field, too, which it would seem should appeal particularly to the pride and interest of the residents of the islands.' Unfortunately, like Wilson, he wrote in vain.

Nevertheless, for reasons that are not entirely clear, the Hawaii Board of Agriculture and Forestry decided to establish a captive flock of Nene in 1927. 'In that year they received a pair from Mr Leighton Hind of Puu Waawaa Ranch and placed the birds under wire at the Territorial Game Farm on Oahu. In the next six years, four additional pairs of adult geese were given the Board by both Mr L. Hind and Mr Herbert Shipman of Hilo.* By 1935 this flock had grown to 42 geese. According to records on file, little difficulty was experienced in the care of the flock.

'In 1935, the Board decided to break up this flock because of fear

* Also of Keaau (Plate 1).

that disease would wipe it out. Accordingly, birds were sent to various sugar and pineapple plantation managers and territorial senators. Ostensibly, the objectives were not only to avert a disaster occurring to the captive birds, but also to stimulate interest in the Nene and to establish many small captive Nene flocks from which breeding stock would always be available should the Board again decide to raise Nene under wire. The plan was a distinct failure. It is suspected that political consideration entered into the decision to distribute the Board's flock about the Territory.

'In 1950, a follow up was made on all of the 42 Nene sent out from the game farm in 1935 and 1936. Only one gander remained available to the Board. This was one of a pair given to the Honolulu Zoo. All the others had either died very soon after their receipt or had been released.

'One wealthy rancher on Molokai who was given four pairs, had a large pen constructed very similar to the Pahakuloa [sic] pen and hired special personnel to care for the geese. The Nene did not lay the first year; so he released them all. This may be the origin of the three Nene observed on Molokai in 1950, although where they had been during the intervening fourteen-year period is a mystery. None of the birds was banded at the game farm before they were shipped out; consequently, the age of the birds given to the rancher was apparently unknown. Since the Hawaiian Goose does not breed until its second or third year, the Molokai flock might well have been composed of first and second year birds. Under those circumstances, production the first season on Molokai in a new environment would have been most unusual. It is a pity that this flock was released to the wild so prematurely and on an island which was never occupied by the Hawaiian Goose, at least in historic times.

'Fourteen geese went to a wealthy senator and were penned at his beach home near Kawaihae, Hawaii. These were retained for several years at sea level with no young produced. At last they, too, were released to the wild. It is to be hoped that this group joined the wild flocks on Hualalai; however, since they were not marked in any way, the valuable knowledge of their movements will never be obtained.

'Six Nene were presented to a wealthy philanthropist who maintained a menagerie on his own private island near Oahu. The history of these geese is absolutely unknown, for after the death of the island's owner, his estate was disposed of and no one seems to know whether the geese were released or given away.

'One other example will suffice. Three geese, one gander and two geese, were sent to the pineapple plantation manager on Lanai.

These were immediately released and were seen for one month thereafter. They were very tame and easily approached. The person who released the birds has good reason to believe that Filipino labourers captured the geese for food, thus, ending this dark chapter in the story of one ill-fated attempt to restore the Nene in Hawaii' (120).

The attitude of Territorial officials toward the native birds, including the Nene, during the 1930s was well expressed by Frank H. Locey (91), then the President of the Board of Commissioners of Agriculture and Forestry of the Territory of Hawaii. 'Things are different in the 1930s than in the 1850s. Today we are strong for any kind of a bird that is either useful, beautiful, or melodius, and that will not injure anything Hawaiian. We now also protect our birds with laws and law-enforcement. Public opinion favours birds today.'

After listing more than 50 different kinds of game birds that had been released in Hawaii, Locey added that 'one impressive factor glares forth from even this brief history of gamebird introduction – failure in introduced birds to become permanently established. Specie after specie [sic] have been imported and disappeared. . . . I want to establish so many birds here that the hunters will not be able, legally, to kill them off.' . . .

'Hawaii can be transformed into a universal aviary, a bird haven where every known species of birds, not injurious to Hawaii, will propagate and thrive under the conditions of their natural habitats. Some day Hawaii will be one grand bird preserve or reservation. We will have here every bird known to the world that will live here to our advantage – and this includes all desirable birds whether game or other birds that beautify our Nature by song or plumage. When that time comes it will not be necessary for people to visit any other part of the world to enjoy or study birds for most of the birds will be living in Hawaii under natural conditions. Hawaii will be one vast Bird Park.

'"Go to Hawaii to see Birds!" will be the popular slogan.'

Although the Board had dispersed its captive Nene flock during 1935 and 1936, as previously mentioned, Mr Locey wrote in 1937 (91) that 'the Board also has interested itself in saving certain native birds, as the *Nene* or Hawaiian Goose, from extinction.' Such published statements over the years constitute virtually the only effort expended to save any rare and endangered Hawaiian plant or animal.

PROPAGATION BEGINS AT POHAKULOA

'High on the volcanic slopes of Mauna Loa and Mt Hualalai, a remnant flock of probably the world's rarest bird ekes out its precarious existence.' Thus did Charles and Elizabeth Schwartz (114) begin their short discussion of the Nene following their field studies during 1946 and 1947. In their published work they urged that an immediate study should be undertaken of this bird's life history and means of survival and, in their unpublished technical report (submitted to both the Board of Commissioners of Agriculture and Forestry in Honolulu, and to the US Fish and Wildlife Service in Washington), they suggested that the State Division of Fish and Game obtain several pairs of Nene in order to initiate a breeding programme with pen-held birds. This suggestion was followed, and it produced far greater results than did their recommendation for a study of the wild birds in their native mountain habitat.

The propagation programme began in 1949, when the Territorial Legislature appropriated $6,000 for a two-year Nene restoration programme. The project itself had been designed by J. Donald Smith, a dedicated game conservationist with the Board of Agriculture and Forestry, who envisioned two interdependent parts of the programme: (1) to rear birds in captivity for eventual release into the native habitat; and (2) to conduct a thorough field study in order to obtain information on the breeding biology of the wild population and the factors which were limiting the success of the species. Smith (120) later published his considered opinion that 'the only effort which is likely to produce lasting and concrete results in conserving the Nene is that directed toward discovering the factors responsible for the Nene's population decline. An ecological study properly conducted of the wild Nene will produce a knowledge of the Nene's life history that no change in attitude of a political agency can affect. It will provide a foundation upon which public service agencies and private organizations can act swiftly and surely either in cooperation or singly to form a Nene conservation program basically sound and continuous in duration and effort.'

Despite the critical importance of the ecological study, however, it was not approved by the Board of Agriculture and Forestry. Moreover, 'the Territorial Government provided no other funds . . . It is to the credit of Mr J. R. Woodworth of the Division of Fish and Game that the project was able to survive for almost nine years on a two-year budget!' (136).

The year 1949 certainly was critical for the Nene. At that time, the world's total population of captive Nene numbered 13 birds: 11 in a flock maintained by Herbert Shipman; a single 'runt' bird in the possession of Mrs Roy Wall at Kealakekua, Hawaii; and a single gander in the Honolulu Zoo. Donald Smith was to estimate the total wild population to be no more than 30 birds in 1951.

Mr Shipman had maintained Nene in captivity on his estate at Keaau (Plate 1), near Hilo, Hawaii, since 1918. Beginning then with only two pairs of geese, the flock had multiplied to 42 birds by 1946. On 1 April of that year, however, when some of the birds were flightless, a tidal wave* deluged his estate at Keaau and killed all

* Actually a Tsunami, caused by underwater volcanic eruptions rather than by tidal activity. This particular wave killed 200 Hawaiians and rendered another 1,300 homeless.

but 11 of the geese. Despite the fiasco of the State's efforts at rearing Nene in the early 1930s, Mr Shipman loaned the State two pairs in August of 1949. These birds were taken to the Forestry and Fish and Game camp at Pohakuloa at an elevation of 1,980 m (6,500 feet) on the Saddle Road of Hawaii, and placed in a specially constructed pen.

Of Mr Shipman's two pairs, however, one goose died of a chronic ailment before nesting was attempted; Mr Shipman* paid the air freight to send the gander of this pair to Slimbridge in March 1951 (Chapter 4). The remaining pair got the project going, albeit at a slow pace. The goose laid four eggs in December 1949. One egg was destroyed accidentally by the goose, two hatched in early January; the fourth egg appeared to be infertile. When this goose completed her clutch of four eggs during the late fall of 1950, the eggs were immediately removed and placed in an incubator. Although the four eggs developed strong-looking embryos, all failed to hatch. Several weeks after the eggs had been taken from the goose, however, she laid a second clutch, this time of five eggs. The goose was allowed to incubate this clutch; all five eggs hatched, and three of the young birds survived to adulthood. Again, during the winter of 1951/1952, the first clutch of this goose's eggs was placed in an incubator, where the eggs failed to hatch. As before, however, the goose laid a second clutch, and the pair raised three young. This original pair continued to raise young as the seasons passed.

A short time after the rearing programme started a pig hunter's dog picked up an uninjured wild female Nene. The gander from the Honolulu Zoo was sent to Pohakuloa as a mate for her. However, this pair produced no fertile eggs over a period of several years, and it was concluded that the gander was sterile. He was returned to the Zoo during the fall of 1953, and a three-year-old gander, that had been raised at Pohakuloa, replaced him. Other shifts in pairing combinations were made. These changes yielded better results, and two of the pairs produced by the original Nene were successful in rearing young.

Both fertility and hatchability of eggs were low during the early years of the programme at Pohakuloa, and only 24 young Nene were raised during the first seven breeding seasons (1949/1950 through 1955/1956). Only 40% of the eggs laid during this period were

*As well as being a pioneer cattleman and rancher, Herbert Cornelius Shipman was a specialist in soil conservation and an expert gardener. He saved several rare orchids and the Mauna Silversword from extinction. After 1950, he visited Slimbridge many times and died in October 1976 aged 84 years.

fertile, and only 53% of the fertile eggs hatched. The programme was falling far short of the hopes of Don Smith, who wrote in 1952 that 'the major objective of this propagation venture is to produce 50 geese a year to be released into the wild to restock the natural range.'

Several men experienced in the problems of raising wild waterfowl in captivity offered suggestions for improving the rearing programme. John Yealland (139; 140), Curator of the Wildfowl Trust, spent several months at Pohakuloa during 1949/1950 and his rearing techniques were used with success. Paul Breese, Director of the Honolulu Zoo and Chairman of the Nene Advisory Committee, also researched new developments in waterfowl care. Wesley Batterson of the Oregon State Game Commission spent three months at the Pohakuloa project during the 1960/1961 breeding season. Other prominent authorities on waterfowl who visited the project were: Jean Delacour, Dillon Ripley, Peter Scott, K. C. Lint, S. T. Johnstone, William J. Sheffler, and William H. Phelps.

Three potential problems were emphasised: inadequate diet, unreliable rearing techniques, and a deficient genetic strain in the Shipman geese.

(a) Diet

In the early stages of the programme, Smith (120) wrote that 'the geese have been fed several different kinds of grains, prepared mashes, and turkey pellets. They have shown a decided preference for whole wheat soaked in water before feeding. They will peck at the mash and seem to take the corn fairly readily, but hardly touch the turkey pellets.

'Other foods that they avidly devour and which are fed daily include pualele, the common sow thistle (*Sonchus oleraceus*). A bed of water cress is maintained by overflow from water storage tanks near the Nene pens, and these greens are fed as an alternate to the pualele. Noteworthy, is the decided preference shown by the geese of all ages from downy young to old doughty ganders for the sow thistle. It is amusing to watch the birds follow the game keeper about the pen anxiously waiting for him to deposit his armload of 'weeds' in their feeders. Quite frequently an armload is devoured by a family group of two adults and three goslings in an hour's time. Baldwin (14) claims this plant was previously considered an important food plant but since the Nene have been largely extirpated from the low-land ranges, it is no longer available to the geese in quantity. It is most abundant around the ranch and farm areas below 4,000 feet and is uncommon in the present main range of the Nene. Curiously,

TABLE 2 Nene Restoration Project Record – 1 July 1949 to 30 June 1978

Nene reared at Pohakuloa		*Nene released Island of Hawaii**						*Nene released Island of Maui*				
Year	*Number*	*Year released*	*Keauhou Sanctuary*	*Keauhou 2 Sanctuary*	*Kahuku Sanctuary*	*Kipuka Ainahou Sanctuary*	*Total*	*Paliku‡*	*Paliku§*	*Hosmer Grove***	*Total*	*Total Nene released*
1949–50	2											
1950–51	3											
1951–52	2											
1952–53	1											
1953–54	4											
1954–55	4											
1955–56	8											
1956–57	12											
1957–58	3											
1958–59	15											
1959–60	17	1960	20	–	–		20	–	–	–	–	20
1960–61	32	1961	11	20	–		31	–	–	–	–	31
1961–62	45	1962	–	35	–		35	30	5	–	35	70
1962–63	54	1963	–	42	–		42	19+(5)	5	–	29	71
1963–64	38	1964	–	–	–		–	20	–	–	28	28
1964–65	41	1965	30	19	–		49	24+(2)	8	–	34	83
1965–66	69	1966	–	–	–		–	–	25	–	25	25
1966–67	84	1967	–	–	75		75	–	–	–	–	75
1967–68	123	1968	–	–	85		85	–	20	–	20	105
1968–69	156	1969	–	33	122		155	50	22	–	72	227

Nene reared at Pohakuloa		Nene released Island of Hawaii*						Nene released Island of Maui				
Year	Number	Year released	Keauhou Sanctuary	Keauhou 2 Sanctuary	Kahuku Sanctuary	Kipuka Ainahou Sanctuary	Total	Paliku‡	Palikus§	Hosmer Grove**	Total	Total Nene released
1969–70	114	1970	106	–	–		106	55	–	–	55	161
1970–71	131	1971	94	–	–		94	–	–	–	–	94
1971–72	104	1972	2	35	–		37	–	44	–	44	81
1972–73	109	1973	13	–	–	61	74	–	50	5	55	129
1973–74	134	1974	–	–	–	123	123	–	–	4	4	127
1974–75	141	1975	–	–	–	135	135	–	–	2	2	137
1975–76	160	1976	–	164	–	–	164	–	34	4	38	202
1976–77	47	1977	–	–	–	–	–	–	47	–	47	47
1977–78	46	1978	–	–	47†	–	47	–	–	1	1	48
Totals	1699		276	348	329	319	1272	205	260	16	489	1761

* All of the Nene released on Hawaii were reared at Pohakuloa.
† 40 young from Pohakuloa; 2 young from Puuanahulu; and 5 adults.
‡ From England.
§ From Pohakuloa.
** From US National Park Service.
( ) Birds from Connecticut.

the pualele is common in the Pohakuloa area at 6,500 feet, which is neither cattle range nor farm land.'

Later, red wheat and several different mash mixtures were imported (or formulated locally), and vitamin and mineral supplements were added to the diet.

(b) Foster mothers

The technique of removing the first clutch of eggs and placing them under a foster brood-parent – a technique that has long been used successfully for other waterfowl – was also adopted. Hence, in addition to electrical incubators, domestic hens (1949–1952), Muscovy ducks (1953–1960), and silky bantams (1950–1965) were used to incubate first clutches of Nene eggs and to act as foster mothers for the goslings. These efforts were generally unsatisfactory, however, because many of the foster mothers failed to become broody during the winter months at Pohakuloa. Consequently, hatchability of fertile eggs remained low.

Experiments begun during the 1965/1966 breeding season revealed that a high percentage of Nene would still lay a second clutch of eggs after incubating and hatching a first one. As soon as they had hatched, the goslings were removed to indoor brooders and the nest was destroyed. Renesting occurred, on average, 31·8 days later and varied between 26 and 39 days. Second clutches were also incubated by the female. As a result of this change in technique, 84 goslings were raised during the 1966/1967 breeding season, the largest number since the beginning of the restoration project. Consequently, the new technique was tried for the entire breeding flock (27 pairs) for the following nesting cycle, though nine of the geese were permitted to lay only one clutch of eggs because of their age, lateness of egg laying, or problems with vicious ganders. A total of 27 first clutches, 18 second clutches, and two third clutches was laid. Thirty breeding pairs were available during the 1967/1968 nesting season, and 123 goslings were raised, bringing the production per pair to 4·1 young. Because of this striking success, ten additional pairs were added to the flock for the 1968/1969 season. The 40 pairs produced 156 young birds. By nearly doubling the production of goslings in a two-year period, the two-man staff at Pohakuloa found, however, that they had created too much work for themselves. Subsequently, therefore, the breeding flock was reduced to 30 pairs, those birds with the best productivity records being selected for the rearing programme. The great increase in the annual production of Nene is shown in Table 2. (Technical data on fertility of eggs, and the weights of eggs, egg shell, goslings, and adults are given in the Appendices.)

(c) The genetic strain

The two pairs of Nene that were loaned to the State by Mr Shipman in 1949 were the result of inbreeding among the offspring of two pairs of geese that he had maintained in captivity since 1918. As the breeding season passed at Pohakuloa, it became evident that there was a high degree of infertility among the inbred ganders of this 40-year-old blood strain, and that the fertility of all eggs laid by the Shipman-strain geese amounted to only 54·5%. Observations of wild Nene nests indicated that their fertility was high.

During March of 1960, a pair of wild adult Nene and one immature bird were captured on the breeding grounds on Mauna Loa and added to the captive flock at Pohakuloa. The adult pair raised many young, which were mated with Shipman-strain birds. The 'new blood' added by the wild birds resulted in a significant increase in the fertility of eggs laid and of young reared (Table 2).

PROPAGATION CONTINUES AT POHAKULOA

In 1956, Paul Breese, then the Director of the Honolulu Zoo and the Chairman of the Nene Advisory Committee, with his characteristic enthusiasm, enlisted the aid of the Conservation Council for Hawaii and other interested citizens to persuade the Territorial Legislature to proclaim the Nene as the 'Official Bird of Hawaii.' This was done in a concurrent resolution in May 1957. The Territorial Government, however, still was not interested in investing funds in the programme to preserve the Nene.

One must doubt that the rearing programme at Pohakuloa would have been continued for many years if Federal Funds had not been awarded to support the project fully (from 1958 to date). Moreover, although Smith (120) had stressed the prime importance of a field study of the biology of the Nene, Dr William Elder's work in the field in Hawaii (September 1956 through August 1957) was funded primarily by external sources, for example, the John Simon Guggenheim Foundation, Yale University and the Bernice P. Bishop Museum (through a Yale-Bishop Museum Fellowship), the International Committee for Bird Preservation, and the McInerny Foundation of Honolulu. In his report, Elder (47) stressed the truism that 'for any species to survive, the rate of production must equal the rate of losses, or *natality* must equal *mortality*. If the balance is tipped ever so slightly, so that mortality exceeds, a species declines. This is what has happened to the Ne-Ne.' As will be pointed out in Chapter 5, however, the weak link in the Nene programme still is a lack of knowledge of the biology of the wild population.

Breese (27) described the efforts involved in obtaining Federal funds for the propagation programme. 'One of the men most closely associated with the Nene program here since its early years is Mr J. R. Woodworth, Chief of the Game Bureau of the State Fish and Game Division. He had direct charge of the Nene program in the early 1950s, and now has overall responsibility for the project through the Nene Ecologist and Nene Propagator who come under his supervision. In 1958 Mr Woodworth personally led the endeavour that obtained Federal financial aid to supplement Hawaii's funds in the Nene Restoration Project. Citing the facts revealed by the ecological study and the positive results achieved by Hawaii, Mr Woodworth obtained the support of the nation's major conservation agencies in introducing a bill into the U.S. Congress. This bill authorized the U.S. Fish and Wildlife Service to spend $15,000 a year for a period of five years to support Hawaii's efforts in carrying on its program of research and management to insure the preservation and to increase the numbers of the Nene in their habitat. This bill was passed by the 85th Congress as Act 891 during the closing days of the session in 1958.

'With the availability of the Federal funds, Mr Richard Griffith, Chief of the Wildlife Division, Region 1, U.S. Fish and Wildlife Service, Portland, Oregon, came to the Islands to survey Hawaii's Nene program and participate with Mr Michio Takata, Division Director, Mr Woodworth, and the Nene personnel of the Fish and Game Division of the State Department of Agriculture and Conservation in drawing up future plans for the continuing Nene Program. These Federal funds helped immeasurably in giving the project financial stability and enabled the Fish and Game Division to expand the Nene program.

'In August of 1960, Mr Ross Leffler, Assistant Secretary of the Interior, and Mr Lansing Parker, Assistant Director, Bureau of Sport Fisheries and Wildlife, made a careful inspection of the rearing project at Pohakuloa as well as the Nene Sanctuary high on Mauna Loa. These officials were so impressed by the success of the Nene Restoration Project that they are in favour of expanding and extending the U.S. Fish and Wildlife Service's support for an additional period.'

And, indeed, not only was the support extended for a second

five-year period but in 1968 the Federal grant to the State was increased to $25,000 per year.

THE POHAKULOA PROGRAMME TODAY

Donald Smith wrote in 1952 that the first Nene pen 'is of two adjacent sections 50′ × 100′, covered top and sides by one-inch mesh fox farm wire. A three-foot wide strip of sheet iron extends around the bottom of the pen on the exterior to prevent mongoose from climbing the wire. Pools 15 feet square, shallow at one end, and 2½ feet at the other were constructed in the centre of each pen. After heavy fertilising and constant watering, a heavy stand of white clover (*Trifolium repens*) and Kikuyu grass (*Pennisetum clandestinum*) was induced to grow on the very sandy and porous soil within the pens' (120).

Additional pens were constructed and changes in design were made over the years as more experience was gained in the propagation programme. In 1973, 31 pens were being used, 21 for breeding purposes and ten as holding pens for immature goslings. The breeding pens measure 7·6 × 7·6 m (25 × 25 ft); holding pens vary from 3·8 × 7·6 m (12·5 × 25 ft) to 7·6 × 15·2 m (25 × 50 ft). The large water pools were removed years ago when it was discovered that they were of no particular value to the Nene (Plate 8). Small, cement water basins about 90 cm (3 ft) in diameter meet all of the Nene's requirements now.

A total of 1,699 Nene were reared at Pohakuloa between the fall of

1949 and May of 1978. Only a few birds (one to four) were produced during the first six years of the programme, but more than 100 birds were reared annually beginning in the 1967/1968 breeding season (Table 2). The work-load for the two man staff proved too heavy and the breeding stock was cut to 32 pairs for the 1972/1973 season, to 12 pairs for 1976/1977 and to six pairs for 1979/1980. The programme presumably will be terminated in a few years (10; 137; 143).

The Nene, as already said, is unique among geese in that it initiates its nesting activities in the fall, when daylengths are decreasing. The dates for first eggs at Pohakuloa (Appendix 6) have ranged from 30 August (1974) to 5 December (1953). New nesting material is placed in the breeding pens every fall; this consists primarily of chopped Bermuda grass and sphagnum moss. The goose makes a scrape in the ground and then lines it with these materials. The nests are usually constructed under redwood lath shelters, but sometimes the geese build beneath the protection of plantings of verbena and wild olive bushes in the pens.

The annual rainfall (about 635 mm or 25 inches) at Pohakuloa is much lower than that in the primary nesting area on the eastern slope of Mauna Loa (about 1,525 mm or 60 inches per year). In order to compensate for the low moisture level, the nests are sprinkled with water twice weekly during the incubation period. The sphagnum moss is of special value because it retains moisture.

Each egg is removed from the nest when laid, and marked with a number and the identification of the goose. Each egg that is removed is replaced by a wooden dummy egg so as not to disturb the sequence. The eggs are then stored until the clutch has been completed, when they are replaced in the nest. Eggs that fail to hatch are examined, and an effort is made to age dead embryos.

The newly hatched goslings are removed from the nest and placed in an indoor brooder. During their first few days in the brooder, they are provided with finely chopped chickweed *Stellaria media*. Some of this food is placed in pans and some is floated in the water fountains. This has proved to be an effective method for getting the goslings started on green food. After a few days, small amounts of 'All' mash-egg-food are sprinkled on the chopped chickweed, thus initiating the young birds to the commercial feed. As the goslings grow older, winter wheat, turnip greens, Siberian kale, watercress, zetra petra rye, pualele and other chopped greens are added to the diet. This assortment of green foods provides the bulk of the diet for the young goslings. They are fed in the morning, at noon, and in the late afternoon.

When approximately two weeks old, the goslings are transferred to an outdoor brooder, and later they are given the freedom of an outdoor run during daylight hours. Although green grass grows luxuriantly in the outdoor runs, the birds prefer pualele, and it is provided daily. The goslings are sexed, banded, and placed in holding pens when between ten and 12 weeks old. Thereafter, they are given the same diet as the adult birds.

A few (3·2%) female Nene lay eggs near the end of their first year, but so far all but one of these eggs have been infertile; almost 84% of the females lay their first eggs during their second year, and about 13% during their third year of life. The greatest period of productivity for two of the oldest geese was between their third and ninth, and third and twelfth years, respectively.

Since the introduction of wild birds to the breeding flock (plus changes in technique), fertility rates have risen to as high as 88%. Fertility of eggs seems to be related primarily to the bloodline of the ganders, the age of the ganders, and the position of the egg in the clutch. Ganders are able to fertilise eggs for at least 13 years, beginning in their first year. An adult, wild-caught gander of unknown age fertilised eggs for eleven years before he died.

Clutch size at Pohakuloa varies from three to six eggs; the average of 182 clutches was 4·26 eggs over a four-year period. Eleven six-egg clutches were laid between 1949 and 1972 (six of these clutches during the 1971/1972 nesting season). Only 35 goslings were successfully reared from these six-egg clutches, but six goslings hatched from one clutch in 1970.

The captive Nene at Pohakuloa have been remarkably free of disease. The death of one bird was attributed to bacterial enteritis; three birds have died of poisoning and necrosis of the liver. Most of the deaths have resulted from accidents (for example, one month-old bird drowned) or from injuries sustained in fighting (two birds were killed because they were so vicious that they could not be used for the propagating programme, or even exhibited in display pens for the public). One goose died from a ruptured oviduct at the time of egg laying.

The oldest goose at Pohakuloa was 16·5 years old at death; another goose lived to an age of 15 years. A gander died at the age of 13 years, and a wild-caught gander of unknown age lived for 11 years at Pohakuloa.

PS

4: The Nene at Slimbridge

It was in the 1930s that Sir Peter Scott* first became aware of the danger threatening the Hawaiian Goose. He wrote to Herbert Shipman, after reading that a small flock was kept in his garden at Keaau, and Mr Shipman promised to donate a pair of geese on condition that Scott came out to Hawaii to fetch them. 'This I might well have done had it not been for the intervention of the Second World War' (117). After the war, when the Wildfowl Trust was established at Slimbridge, a letter was written to the Government of the Territory of Hawaii asking what steps they were proposing to take to save the remnant from extinction, 'but the interventions of a well-meaning society on the opposite side of the globe are unlikely to be well received by Government officials in any part of the world. Our letter remained unanswered.' A little later, Scott was in correspondence on another subject with the American scientist, Charles Schwartz, who mentioned that he had been invited to visit Hawaii in order to advise on game bird management. He added that he was very uncertain about whether he would take the job. Scott wrote at once imploring Schwartz to accept, on the grounds that this was perhaps the only opportunity of saving the Hawaiian Goose. 'Whether my importunity had any effect I do not know, but Dr Schwartz took the job and focused the attention of the Department on this tiny remnant population so successfully that a new project was planned for the breeding of Hawaiian Geese in captivity. In charge of this project was an ornithologist, Don Smith, who had studied for some time at the Delta Waterfowl Research Station. When he sought advice from

* Peter Scott was knighted in 1970 for his service to conservation and the environment. In this field his contribution has been unique: he is one of the Founders of the World Wildlife Fund and Chairman of the Survival Service Commission of the International Union for the Conservation of Nature; he is also, of course, the Founder (in 1946) and Director of the Wildfowl Trust. His father, Captain R. F. Scott of the Antarctic, wrote to his mother 'Make the boy interested in Natural History,' and Peter Scott himself says that he cannot remember a time when he was *not* interested in the subject (117). But he is much more than just a very successful naturalist. He is a talented artist, writer and broadcaster, a distinguished naval officer, a British gliding champion, and an Olympic yachtsman.

Delta, Al Hochbaum in turn suggested that he might consult the Wildfowl Trust, and in this way we found ourselves invited to help with the Pohakuloa Propagation Project' (117).

In the following year, this help took the form of sending John Yealland, the Curator of the Wildfowl Trust, to Hawaii to inaugurate, during the first season at Pohakuloa, the standard rearing programme used at Slimbridge. On 3 May 1950, he returned with two Nene. These birds had been presented by Herbert Shipman and were part of his small captive flock from which had also come the two pairs lent to the Pohakuloa project (Plate 1). It was hoped that by sending a pair of geese to the Wildfowl Trust, the chances of saving the species by captive breeding would be greater and that, eventually, it might be possible to return young Nene to the wild.

In March 1951, to everyone's surprise and embarrassment, both birds went under a hedgerow, built nests and laid eggs. A cable was sent to Hawaii, and the eggs removed in order to encourage a second clutch.* After only seven days, a fine gander arrived (116). He was one of the birds lent to the Pohakuloa project, and Mr Shipman's generosity in agreeing to send him to Slimbridge instead, and Don Smith's promptness in arranging transport, were much appreciated. The female geese also were delighted. However, the gander was already in full moult (Plate 20) for, so far as he was concerned, the breeding season was over; so their second clutches were again infertile. The three birds were kept together in a spacious pen and given the names of Hawaiian royalty: Kamehameha for the gander and Emma and Kaiulani for his two wives.†

On 18 February 1952, Emma laid the first egg of her second season. She produced a total of four and, before the clutch was complete, Kaiulani had laid three of her eventual five. Although the gander divided his attention unequally, preferring to be with Emma even while Kaiulani was laying, both clutches contained only one infertile egg. The eggs were taken away and incubated under broody hens, and two of Emma's and three of Kaiulani's hatched. Both second clutches contained five eggs; four of Emma's hatched, but Kaiulani, who had been left to care for her own eggs, did not sit well and allowed them to chill. Thus 19 eggs were laid, 14 fertilised and nine hatched (Figure 25), all of which were safely raised.

*The infertile eggs were blown and the Scott household enjoyed an omelette from the second rarest bird in the world.

† The first Kamehameha declared himself King of the Islands in 1795, 17 years after the death of Captain Cook, and reigned until 1819. Queen Emma was the consort of Kamehameha IV (1854–63), and Princess Kaiulani was heir to the throne at the time that the Hawaiian Monarchy was abolished in 1893.

Fig. 25. Production of Nene at Slimbridge in 1952. Emma laid first, on 18 February, and of the two goslings that hatched, one was 'cottony' (see Plate 22). Drawing by John Turner *from one by* Peter Scott *in* Ann. Rep. Wildfowl Trust 5:54.

In the following year, and in 1954 and 1955, only four goslings were reared (116). These disappointing results were due to the ill health of both females; Kaiulani suffered a prolapse, but was skillfully repaired and continued to lay until 1959. She finally died in March 1960, having produced a total of 99 eggs and 11 goslings. Her skin is now stuffed and mounted in the museum at Slimbridge. Emma had preen gland trouble and died in January 1954, when a post-mortem examination showed that her adrenal glands had been diseased. Kamehameha himself survived until the summer of 1963, by which time he was more than 14 years old and blind; but he had had five wives, and 49 of his descendants had been returned to Hawaii for release into the wild (Table 2 and Appendix 6).

HUSBANDRY

Pairs and trios of Hawaiian Geese at Slimbridge are usually put in grassy pens of at least 230 m² (300 sq yd), in which they are visually isolated, as far as possible, from others of their kind. Because there is no danger to the adults from predators (such as the mongooses at Pohakuloa), the enclosures do not need to be covered. Recently, up to 35 young pairs have been placed in a much larger enclosure of 5·5 ha (14 acres) and left to space themselves out. The arrangement has been a considerable success: nesting sites have been selected at about 45 m (50 yd) apart, and most pairs have attempted to breed. There is a pool with running water in every pen, although the birds do not use it much except for drinking and a rare bathe. Food consists of grass, whatever other plants grow in the lawn, and wheat grain, with the addition of poultry layers' pellets from the beginning of January until the end of the breeding season. A plentiful supply of quartz and shell grit is available at all times.

Because the weather is not warm when the birds start to lay, they prefer nesting sites that offer some protection from the wind and rain. At first, straw huts were used to shelter them at night in winter, which was probably unnecessary as they are quite hardy; but the huts did serve another purpose in that the birds frequently nested in them (72). The construction was a hazard, because straw is particularly likely to become mouldy and infect sitting adults and eggs with a fungal disease, aspergillosis, so 'wigwams' were offered instead. These are woven from willow stocks pushed into the ground on the circumference of a circle of 0·6 m (2 ft) across and tied at 1·0 m (3 ft) from the ground (Plate 12). As with many non-domestic birds, Nene are reluctant to try laying again in a nest that has been 'robbed'. So, if clutches are removed, alternative sites are necessary (82). On the other hand, if the goose is left to hatch her own eggs, then the same site is used the following season (typical also of wild waterfowl).

The small pens at Slimbridge usually contain at least three potential nest sites for the pair to choose from. These may be a wigwam, a hollowed tree trunk, a large dog-kennel, a hollowed space beneath a bush, two hurdles tied together like a tent, or a single hurdle leant against the fence. These are, to some extent, varied every season in order to persuade the birds that all the sites are new. Alternatively the pair can be moved to another pen with completely different opportunities. A little nesting material is added, usually wood shavings or dead leaves: straw and hay are avoided because of the risk of moulds already mentioned.

The wigwams are probably the preferred sites, but Nene also like to nest beside the fence of their pens, against a shrub or a clump of grass. The 'tents', boxes and hollowed trees are unpopular, unless the ceiling is high enough for the bird to enter without stooping.

Nene frequently choose to build in the highest ground available, such as a bank, perhaps because such sites are drier. Unlike captive Canadas and many other goose species, they seldom select the small islands in their pools. Presumably in the course of thousands of years' evolution in a habitat with no permanent marshes, lakes, or streams, and few ground predators, any tendency to prefer a nesting site surrounded by water has been lost.

As eggs are laid (usually in the early morning on alternate days), they are collected and replaced by wooden replicas. This is because the birds do not guard the nest constantly until the clutch is complete, and prolonged freezing temperatures and egg-eating crows are potential threats. Eggs are stored at about 5°C (40°F) until the females start to incubate, the dummy eggs and nesting material (leaves and any down that the goose has pulled) are then removed, and the real eggs put under a broody domestic hen. Recently, the technique has been to leave the eggs under the hen for only 14 days, and to use an artificial incubator for the second two weeks, and for hatching. This has been done mainly to eliminate the possibility of disease being transmitted to the young Nene by the hens at hatching and during rearing. Goslings are now usually reared in small groups without any foster mother.

PRODUCTION AND MORTALITY

By the end of 1958, a flock of 53 Nene had been built up at Slimbridge, and nine pairs loaned to other collections. Eight years later, 195 Nene had been reared at Slimbridge and 124 elsewhere (72). Between 1952 and 1973, 754 Nene are known to have fledged in Europe, and at this point it was felt that they were so numerous (there were 232 at Slimbridge and 200 on loan to 16 establishments in the UK, ten in the USA and nine in continental Europe) that, with the full agreement of the US Department of the Interior, Fish and Wildlife Service, it was possible to offer some for sale. It had been decided that the captive breeding programme in Hawaii was fully able to meet demands for birds for release into the wild, so that the Wildfowl Trust could distribute its birds freely to any zoos or waterfowl breeders that wanted them. Obviously it then became more difficult to follow the increase in numbers of Slimbridge-stock birds, but as detail in Appendix 6 shows, staff at Slimbridge have raised 782 birds in 28 years.* The most successful breeder to have

* The Wildfowl Trust as a whole (including the Peakirk, Martin Mere, Washington and Arundel branches) has produced 888 Nene in this time.

Wildfowl Trust geese on loan was Mr Jack Williams of Norwich. From only a few pairs, he reared 213 between 1963 and 1977, and of the 200 returned to Hawaii from England, 50 were produced by him*. Other particularly successful establishments have been the Peakirk and Washington refuges of the Wildfowl Trust. Leckford in Hampshire (where 70 were reared by Mr Terry Jones), Bentley in Sussex, and West Berlin Zoo.

*Jack Williams, a superb aviculturalist, died in July 1979. His best Nene year was 1970, when all his four females (on loan from Slimbridge) laid three clutches, and from 56 eggs, 33 goslings were reared. But even more remarkable, perhaps, was that in 1968 he had 31 eggs from only two pairs and reared 25 of their goslings. In 1975, six goslings were produced by a pair of which the gander was 16 years old, within a few months of the maximum age recorded at Slimbridge or Pohakuloa.

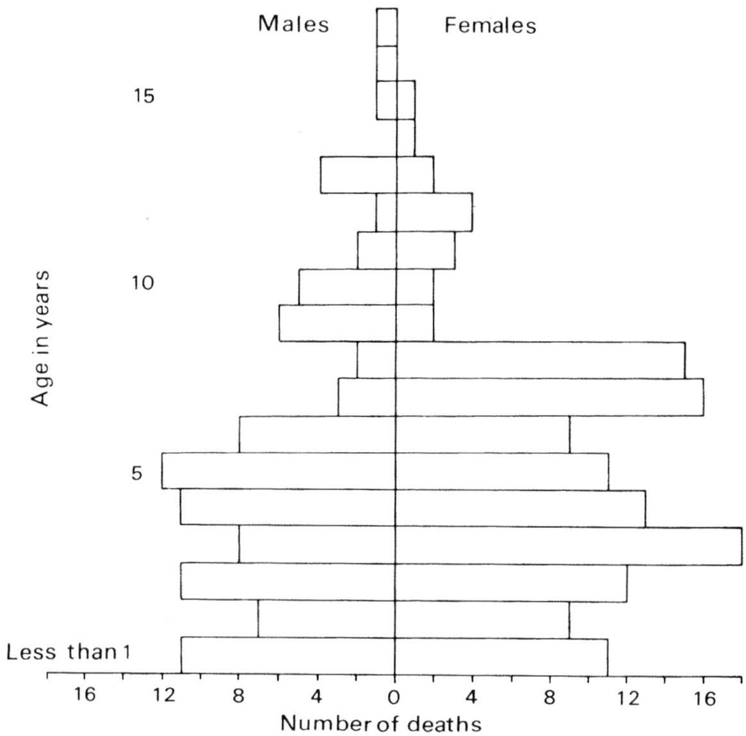

Fig. 26. 223 Slimbridge Nene deaths plotted against the birds' age – 94 males are on the left and 129 females on the right. Females tend to die younger than the ganders, and most females are dead before they are nine years old. Only birds dying after they were three months old are included in this and Fig. 27. See Appendix 7 for details.

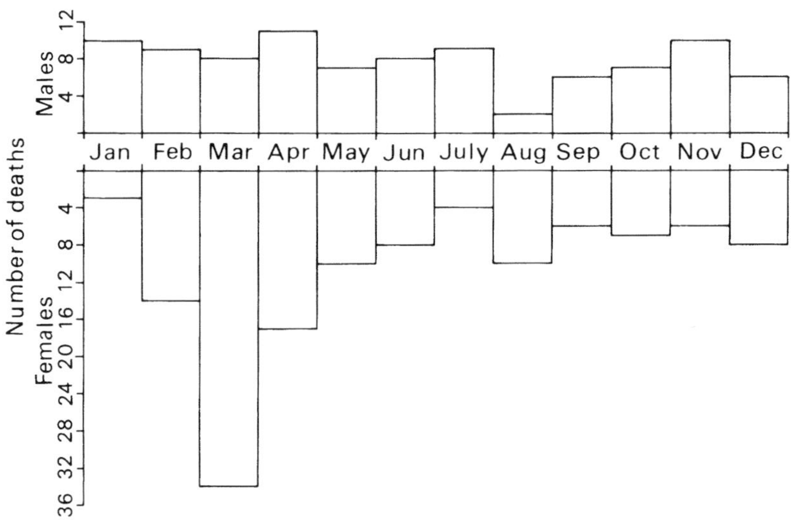

Fig. 27. Slimbridge Nene deaths in relation to the month of the year. The 93 ganders (above) show no noteable pattern, whereas 127 females (below) are seen to die mainly during the breeding season of February, March and April. See Appendix 7 for details.

Of those ringed Nene whose sex was known, 385 were males and 373 females. One hundred and fifty-eight juveniles died, either at hatching or before they reached three months of age when they are ringed, and those comprised 73 males and 85 females. All the evidence, therefore, suggests that the sex ratio is equal, with females having a slightly higher death rate as goslings. This is what one would expect from a knowledge of other waterfowl: the female is more prone to die at hatching (75) and at most stages after that. In ducks, for instance, the adult population is often composed of a preponderance of drakes (15). At Slimbridge, the majority of Nene surviving at nine years are ganders (Figure 26), and females have a significantly higher death rate up to that age.

During the first 25 years of the Slimbridge 'programme', 269 birds are known to have died after three months of age; 44 succumbed to injuries of some kind and 225 to more natural causes. Fatal injury is slightly more likely to affect the ganders than the geese (24 ganders:20 geese), perhaps because it is the males that pick the fights with swans and larger geese who finally do the damage, or because the males actually place themselves between an aggressor and their wives, and are the first to be killed. Females are more prone to die 'naturally' (96 gander:129 geese), and it is particularly interesting to look at the relationship between their deaths and the month of the year (Figure 27). Clearly, females are especially likely to die during the breeding season. This tendency is apparent in many goose species – egg-laying and incubation are times of stress for any female – but is particularly marked in captive Nene in Europe. At post-mortem, they are often found to have layers of fat around the gut and oviduct which must interfere to some extent with egg-laying. In order to 'correct this condition', a low-calorie diet was tried at Slimbridge during one year, but with little effect on mortality. The storage of fat may well be a normal response to the winter cold of England. Perhaps if other geese could be induced to lay in February, they too would suffer from egg-impaction due to fat around the oviduct (72). Fortunately for them, they usually lay much later, in the warmer weather of spring.

Male Nene are slightly less likely to die during the late summer, but in general, they show a much more even, year-round, mortality pattern than the females.

The most common findings at post-mortem examination are bacterial disease (including avian tuberculosis), aspergillosis, complications of egg-laying (such as the impaction of the oviduct already described, and peritonitis), heavy parasite infestation especially of intestinal worms (11), and degenerative diseases such as hardening of the arteries (see Appendix 7 for details). Nearly half the Nene dying during the last ten years have shown hardened arteries (atherosclerosis), especially the older birds (Figure 33). A particularly interesting condition of Nene (and of no other waterfowl) is something that closely resembles avian pox (Plate 23). It affects the feet and the bill and, at an advanced stage, is seriously incapacitating (83). Hawaiian bird species are thought to be peculiarly susceptible to pox (130), and the introduction of blood-sucking mosquitoes to the islands seems to have been responsible for the extinction of many of them (see Chapter 6).

The average age at death of Slimbridge-ringed birds is 4·8 years (5·0 for males and 4·6 for females). The maximum age of a Slimbridge-bred bird (a gander) at death was 17·7 years (as compared with 16·5 years at Pohakuloa). This life expectancy seems rather short for a goose (Johnsgard (67) gives captivity records for Canada Geese at 33 years and the Greylag at 26 years) and contrasts poorly with the 42 years claimed for a Nene gander that died at

Clères, France (39). Possible reasons for the mortality pattern at Slimbridge are discussed later.

THE BREEDING SEASON

No Slimbridge females have laid at one year old. Fifty-six per cent lay at the end of their second, 30% lay for the first time at three years of age, 2% at four years and 4% at five years. A few (8%) never produce eggs. These data (which can be compared with those from Pohakuloa in the last Chapter) do not represent a true picture of sexual maturity in the Slimbridge birds, since health and human interference are also important. Some of the 14% of three-year-olds not laying will be diseased, and others will be poorly mated to the gander they have been given. He must display in order to encourage his female to lay, and ill health and personal preference will affect whether he does so or not.*

Apart from the Cape Barren Goose *Cereopsis* of Australia, the Nene is the first goose to breed in the English spring. Daylengths of 8·8 hours (24 January) are required before egg-laying ever starts at Slimbridge (which is on a latitude of about 52°N), and 9·5 hours is more usual, 8 February being the 'average' date of first Nene eggs over 29 years (Appendix 6). Live sperm have been found in a semen sample taken from one male as late as 25 April (most unpaired Nene ganders at Slimbridge produce sperm from the beginning of February to the middle of March – see Figure 28), and the 'average' date for starting the last replacement clutch of the year is 10 April (13·8 hours of daylight). In other words, if all clutches are included, most Nene eggs are laid between the sixth and the fourteenth weeks of the year. Since eggs are not laid through the late spring and summer, it is likely that long days engage some kind of 'switch-off' mechanism. It was found, for instance, that in an enclosure at Slimbridge where floodlights stayed on until midnight in winter, Nene moulted three months early in January, and failed to breed at all.

In Hawaii, at a latitude of 20°N, sunrise and sunset are never

* The only permanent 'misalliance' between a Nene and another species at Slimbridge involved a male Hawaiian Goose and a female *Cereopsis* (84). The pair-bond lasted for three years, until she died. The female laid every spring (after the birds had copulated on land, which is normal in both species), but the eggs were always infertile. Hybridisation with any wild goose species would be almost impossible, as their breeding seasons do not overlap. However, in Hawaii, hybrids are reported between Nene and farmyard Chinese Geese *Anser cygnoides* (60) which, of course, have been selectively bred for a long laying season. Skins of such hybrids are preserved in the Bishop Museum, Honolulu.

Fig. 28. Nene sperm, greatly enlarged.

separated by more than 13·5 hours or by less than 10·8 hours. So it is conceivable that in its native haunts the Nene could breed throughout the year and, in fact, only during May, June and July are eggs not found. Henshaw (62) was told of nests (presumably in the lowlands, where Nene no longer occur) at the end of August, Pohakuloa has had its earliest egg then (Appendix 6), and at the other extreme, a wild nest was seen in the Keauhou Sanctuary on 10 April, which had hatched by the 15th. Possibly in the days when a sizeable population of geese bred in the lowlands, they had an even longer breeding season than they do now. The Nene's failure to breed in England at the correct autumn daylength – when birds gain weight (Figure 10), pairs display and even copulate – might be explained in this way: long summer days inhibit breeding and are followed by only a short period in September before daylengths decline *below* the threshold of stimulation. So, although hormones start being secreted, and lead to the appropriate displays, the neuroendocrine mechanism soon becomes inhibited with the onset of shorter days in late October. Breeding behaviour is not again stimulated until late January, when the 'right' daylength comes round once more (101).

The date varies on which a particular female will lay. The first time she lays she will do so, on average, eight days later than in subsequent years. Experience is more *directly* relevant than is age: a four-year-old nesting for the first time will lay later than other four-year-olds that bred in previous years and will lay earlier herself when she is five (assuming that she keeps the same mate, and is not moved to a different latitude or climate). The earliest egg at Slimbridge was laid on 24 January 1961 by a three-year-old, who had first bred in March 1960. Cold weather is probably responsible for most of the variation in laying date among established breeders at a particular site. The latest 'first egg date' at Slimbridge was 27 February 1963, after one of the severest winters recorded this century.

CLUTCH AND EGG SIZE

The most common clutches at Slimbridge contain four or five eggs, the normal range is from three to six, and the average number of eggs in 600 clutches is 4·28. This excludes clutches of two, and dropped single eggs laid by sick or very old birds – these are thought

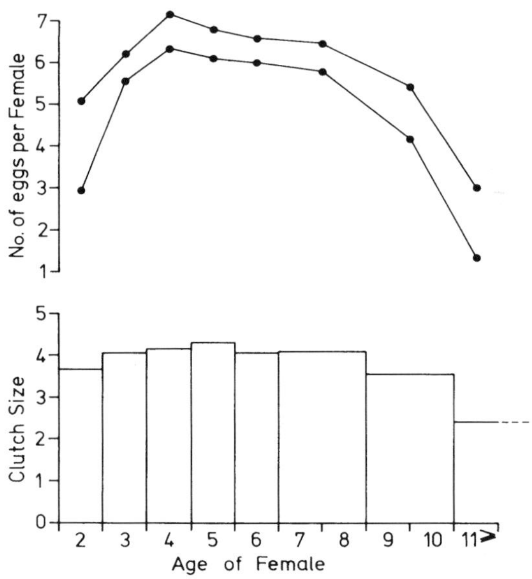

Fig. 29. Number of eggs in 912 Slimbridge Nene clutches plotted against the age of the laying female. Many females lay repeat clutches, and on the graphs above are shown the total number of eggs laid in any season by: A the paired, or available, breeding females and; L those females that actually laid.

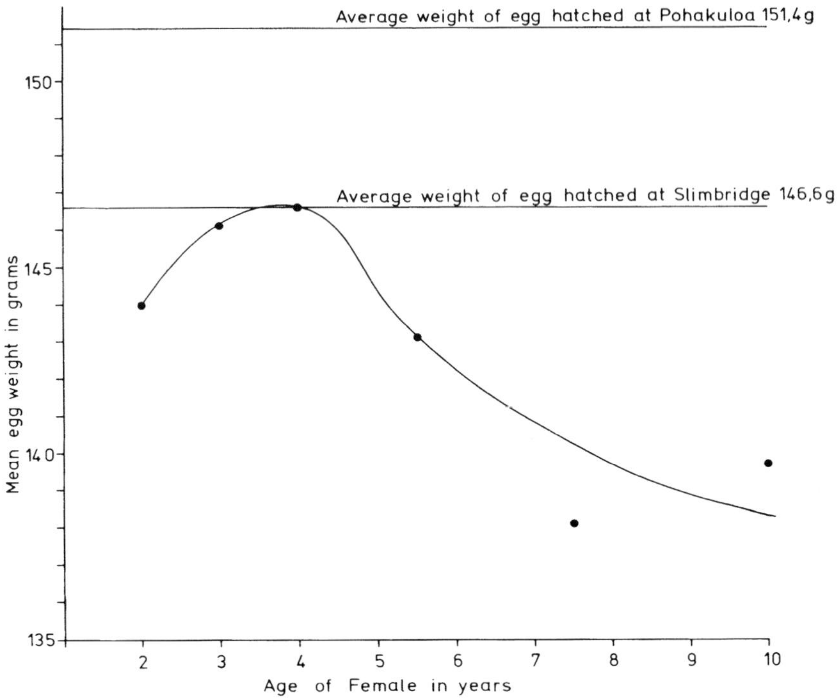

Fig. 30. The average weight of 619 Slimbridge Nene eggs, in grams, plotted against the age in years of the laying females. Four-year-old birds tend to lay the heaviest eggs. See Appendix 4 for details.

to be abnormal in the sense that the female seems never to make a proper nest and attempt to incubate a clutch containing, initially, less than three eggs. Two-year-old females typically lay smaller clutches than three, four and five-year-olds. At eight or nine years, production has dropped to the early level again. Thereafter, clutch size declines sharply, until the typical pattern for a 12-year-old is to lay a single egg, often small and thin-shelled. Forty-five per cent of birds lay replacement clutches, on average 18 days after their eggs are removed, and 7% produce third clutches. Again, the 'middle-aged' geese are more likely than younger or older birds to lay for a second time.

Egg weight has been briefly discussed already. The normal range at Slimbridge is 88·0–173·0 g, and the average is 144·0 g. Large numbers (700) have been weighed and it has been found that they vary in size between certain categories. Thus, those eggs that hatch are significantly heavier than infertile ones, first eggs in a clutch are significantly lighter than subsequent ones (both these features are found at Pohakuloa too), and eggs of the first clutch weigh slightly more on average than those of the second or third. Eggs also at first increase in weight with the age of the mother – the heaviest eggs are laid by three and four-year-olds (Figure 30) – and then decline significantly in weight as the birds grow older (see Appendix 4 for details).

FERTILITY

In the first years, the fertility of the Nene at Slimbridge was fairly good – fourteen out of nineteen eggs (74%) were fertilised in 1952 (Figure 25) but the ganders' performance soon declined (72). By the early 1960s, the fertility of first and second clutches was averaging a low 41%, and that of third clutches was only 23%. The position of the egg in the laying sequence did not affect the likelihood of

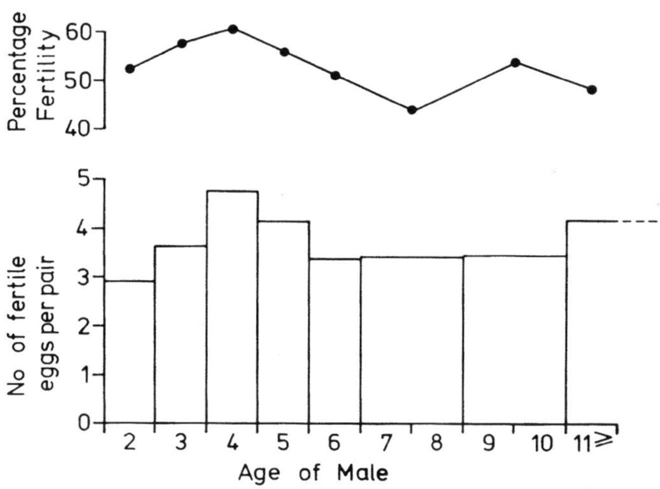

Fig. 31. The histogram shows the number of fertile eggs produced by paired Nenes at Slimbridge plotted against the age of the gander. Four-year-olds tend to have the highest number of fertile eggs, although surviving older males can continue to be quite productive. Percentage fertility (that is, the number of fertile eggs compared with the number available to fertilise) is shown in the graph at the top. See Appendix 6 for details.

sterility – first eggs were as prone to be as 'clear' as last ones (unlike the situation reported from Pohakuloa). The ganders fertilised most eggs between their fourth and eighth years of age, and four to six-year-olds (again, the 'middle-aged' birds) were particularly valuable (Figure 31).

It seems likely that the problem of low fertility arose from inbreeding, since all the Slimbridge stock stemmed from three birds, and these came from the Shipman flock that was already inbred. An introduction of new genetic material was felt to be essential. So, in September 1962, as already mentioned, two young wild-blood males from Pohakuloa were sent to Slimbridge. This out-breeding was a considerable success: by 1967, fertility was 57% and by 1968, 63% (see Appendix 6). Two more young ganders arrived from Hawaii in June 1967; by 1971 fertility was well over 70%, and has since remained fairly high.

Vigorous selection of the males that were available for breeding also helped to increase the production of goslings. Ganders known to be fertile, or to have had particularly fertile fathers, were paired (and sometimes trioed) with geese known to lay a large number of eggs each season, or with their daughters. Samples of semen from doubtful males were collected and examined microscopically (64). Any Nene found to produce deformed sperm, or relatively few sperm, or sperm that showed little movement, were eliminated from the breeding programme. These males continue to lead apparently contented, if somewhat dissipated, lives hanging around the restaurant at Slimbridge, where their shameless begging for titbits of bread and cake endears them to the public.

It is notable that inbreeding did not affect the fecundity of the females. Clutch size in captivity is possibly even higher than in the wild, and females of pure 'Shipman' descent sometimes produce 14 or 15 eggs in a season.

HATCHING AND GROWTH

During incubation, eggs are candled regularly and infertile ones removed. Apart from the large number of 'clear' eggs, the embryo in some of the fertilised eggs either dies early when the egg is invaded by bacteria, or the gosling dies during hatching. Addled (rotten) eggs have made up nearly 12% of those produced every year, and 5% have been 'dead-in-shell' (see Appendix 6). The number of these wasted eggs remains constant through the life of the goose and cannot be related to her age, unlike fertility, clutch size, and egg weight. This

Fig. 32. The average number of Nene goslings produced at Slimbridge plotted against the age of 103 laying females who produced at least one gosling during their lives. As expected, the most productive mothers are four- or five-year-olds.

perhaps suggests that the artificial incubation technique is at fault. First clutches at Slimbridge have, until recently, been incubated entirely by domestic hens which also fostered the goslings, but this system has now changed. The climate in February is too unreliable to allow the geese themselves to incubate, as the occasional prolonged frost can cause desertion. Even second clutches are not always safe, and only the eggs laid late in the season are left with the females.

The number of goslings produced every season can be correlated with the age of the mother, and four and five-year-olds have most offspring (Figure 32). This is not unexpected since the number of fertile eggs increases in the same way, while addled and dead-in-shell eggs do not. The slight initial increase in egg weight with age may also be important here, as it seems that larger eggs are more likely to hatch. On average, 81% of hatched goslings are raised successfully (72).

The range in weight at hatching of Slimbridge goslings (see Appendix 5 for details) is considerable (63·5–115·5 g) and, in general, the smaller ones (more than 15 g under the average of 94 g) do not survive the first days of life. Food consists of chick crumbs, with sieved egg yolk scattered on top (the young bird shows a preference for yellow and green (74)), and fresh greenfood, espe-

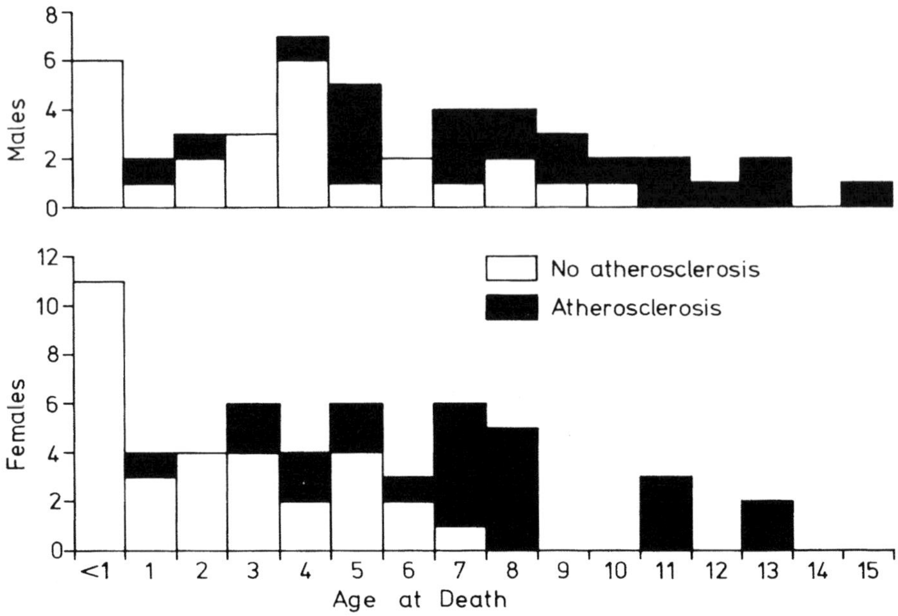

Fig. 33. One hundred and eleven dead Nene at Slimbridge have been examined recently for lesions of atherosclerosis (hardened arteries). Forty-one per cent of both sexes (47 ganders are shown at the top) had lesions, and almost all the older birds were affected.

cially watercress and dandelion leaves.* By the third day, Nene goslings are feeding well and gaining weight (Figure 34), and at three weeks old they are about eight times their hatching weight (this is less than in typical arctic geese, which may be as much as 14 times their hatching weight at three weeks of age (78)). Their scapular and flank feathers are now just starting to show. At five weeks, the quills on their wings are visible (Plate 14), and at 8½–9 weeks the birds can fly. This is two to three weeks earlier than in Hawaii (47), and perhaps associated with this more rapid growth rate is the relatively short life expectancy of Nene at Slimbridge. The hardening-of-the-arteries (atherosclerosis) shown by just under half of the Slimbridge Nene at post-mortem examination (Figure 33), may be related to their early, fast growth rate, associated with a rich

* It is interesting that the Nene goslings' liking for watercress (noted also at Pohakuloa) is not shared by other geese reared at Slimbridge.

(high protein) diet and a rather low level of activity. In addition, the longer daylight hours during spring in Europe may mean that the birds feed for a longer period of every 24 hours, maintaining a higher metabolic rate, and so mature faster. Blaauw (22) noted that his captive birds in Holland grew very fast and 'at the age of about nine weeks the wings have to be cut to prevent them from flying away'. Since high protein commercial rations were not available then, it is possible that climate and daylength in the European spring are more important.

A naturally slow-maturing goose, fed a rich diet, often with a shortage of green vegetation in the winter, and leading a rather sedentary existence (until recently no Slimbridge Nene was left full-winged), might be expected to show degenerative diseases

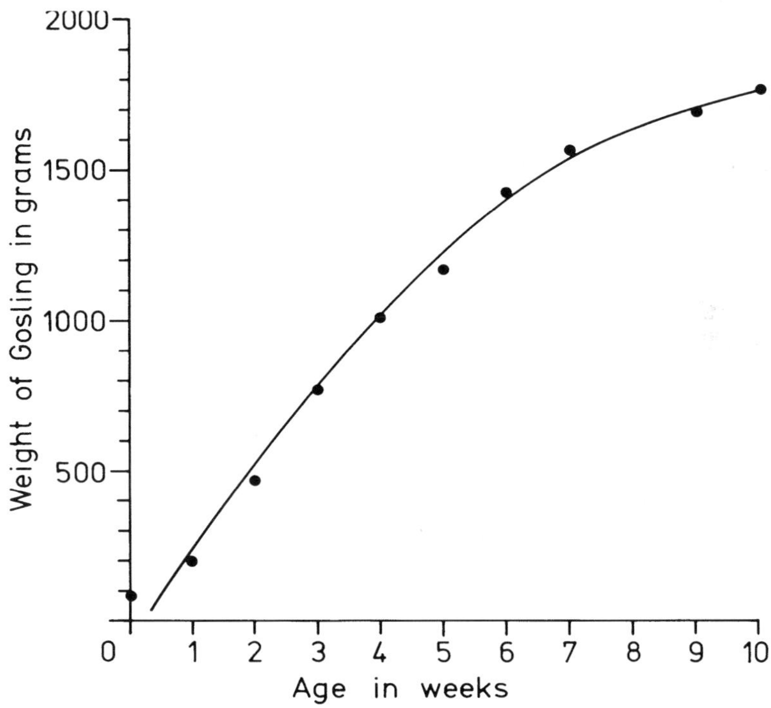

Fig. 34. The growth curve of Nene goslings at Slimbridge. See Appendix 5 for details.

early. These may, in turn, reduce the expectation of life. During 1973, for the first time, 29 selected Nene, mainly ganders, were allowed to fly around the Wildfowl Trust grounds at Slimbridge. The experiment will investigate their tendency to wander, and test the effect of increased exercise upon fertility and ageing.

SUMMARY OF THE DIFFERENCE BETWEEN THE TWO MAIN CAPTIVE FLOCKS

It is clear that Nene at Pohakuloa and Slimbridge differ in a number of respects. First, the breeding season in England starts about five months later (Appendix 6), and is much shorter. As already explained, this is almost certainly due to the fact that English daylengths at high summer *and* in the depths of winter are incapable of stimulating the correct hormone production, whereas winter days in Hawaii never shorten sufficiently to be inhibitory, and birds may remain in breeding condition there for many months. In addition, as correct daylengths last longer in Hawaii, repeat clutches are commoner and a few captive Pohakuloan females (ten since 1953) have laid four times in a single season. Egg-laying dates at latitudes between Hawaii and Slimbridge are of interest here. At Patuxent (38°N) in Maryland, for instance, eggs are laid as early as 1 January, and in Connecticut at 42°N, on 29 December. A Shipman-strain female kept at Nashville, Tennessee (36°N), laid four clutches (which seems to be the only record outside Hawaii) totalling 19 eggs when she was four years old.

Secondly, eggs at Pohakuloa are on average about 5 g heavier than those laid at Slimbridge (see Appendix 4). The size of eggs in a number of birds is known to be influenced by the weather, lower temperatures meaning that smaller ones are laid (C. M. Perrins pers. com.). The night temperature during February at Slimbridge can be below freezing, but so it may be at Pohakuloa (16). The great difference between the two places is in the temperature range: during a winter day in Hawaii, the thermometer almost always rises above 16°C (61°F). In February 1972, the high and low temperatures at Pohakuloa were 26° and −5°C (78° and 23°F), and in 1973 were 28° and −8°C (82° and 17°F). The equivalents at Slimbridge were 10° and −1°C, (50° and 30°F) and 12° and −6°C (54° and 21°F). Thus the much colder days at Slimbridge may be the reason why Nene eggs laid there are lighter. Egg weight in turn affects the weight of the goslings, and Slimbridge Nene are on average 8 g smaller at hatching than those at Pohakuloa (see Appendix 5).

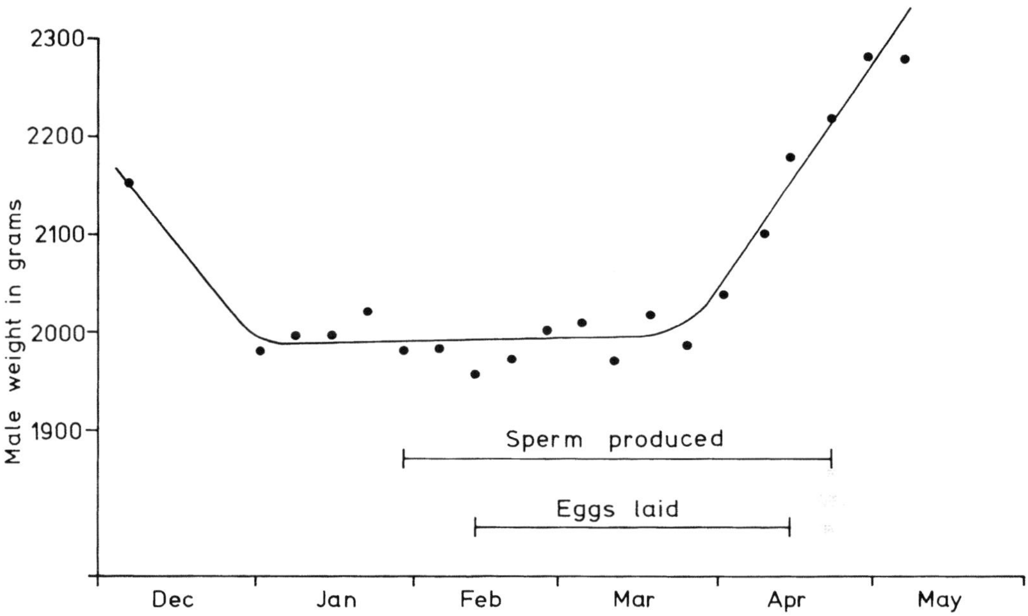

Fig. 35. The average weight of 8 Slimbridge Nene males, handled at weekly intervals through the winter and early spring for examination of their semen. The period during which live sperms were obtained is indicated, as also is the period of egg-laying in the rest of the Slimbridge population.

Thirdly, although lighter at the start, Slimbridge Nene grow (or at least fledge) faster than at Pohakuloa, and certainly are no smaller as adults. Their weight changes at Slimbridge have already been illustrated in Chapter 2 (Figure 10). Apart from the breeding months of February and March, when few Nene have been handled, there are plenty of English data from which to speculate, although little comparative material from Hawaii (see Appendix 2). Most geese are at their heaviest just before the breeding season: the male can then 'afford' to spend time defending his nest and territory, rather than feeding, and the female can produce her eggs and be relatively inactive while sitting on them for a month or so. Slimbridge Nene seem unusual in that both sexes are relatively light in January at the beginning of their breeding season, and the few males that have been weighed through February and March did not start to put on weight until half way through the second month, by which time the grass had started to grow lush, and the egg-laying season was almost over (Figure 35). Perhaps Slimbridge Nene are physiologically 'expecting' to breed from October on, when they increase in weight as they do in Hawaii, but daylength problems prevent them from doing so, and they lose condition as winter approaches. Wing moult in Slimbridge Nene occurs later, and is less variable in its timing than it is in Hawaii. Insufficient research has been done to show what particular daylength (if any) is required to trigger moult: 14 hours of light certainly seems to cause Slimbridge Nene to lose their flight feathers. But if a day that long were actually *needed*, then the birds in Hawaii could not be expected to become flightless at all!

Direct comparison of mortality patterns at Slimbridge and Pohakuloa is difficult, partly because the data from Hawaii are not detailed. In any case, 85% of the young Pohakuloan birds have been released into the wild while only 20% of English Nene have been, so that the relative numbers remaining in captivity to become old and die are very dissimilar. However, captive females in Hawaii lay at an earlier age, despite growing more slowly, and yet produce their most fertile eggs in their ninth to eleventh years – considerably later than in England. The decline in the number of eggs laid in a season is also not apparent until the bird is ten years old, two years later than at Slimbridge. It is likely that Pohakuloa Nene on average live longer than Slimbridge ones – and they do not regularly suffer from oviduct troubles at laying (another probable temperature effect). It is interesting that the *maximum* life span at the two places is similar.

The success recently achieved at Pohakuloa in improving incubation by allowing Nene to hatch their own eggs cannot occur at Slimbridge. Again, the English climate is to blame, but the cold days and nights are not the only factor. As captive female Nene take 32 days on average between hatching their first brood and re-laying, few second clutches would be produced in England if birds were left on their first eggs – the 'season' simply is not long enough. Another advantage of this Pohakuloa system is that additional nesting sites seem not to be essential. The bird apparently assesses the safety of a site on whether eggs hatch there – even when the young are lost soon afterwards. New incubation techniques are being researched at Slimbridge, through an investigation of the natural nest, in an effort to reduce the number of addled and dead-in-shell eggs. A fibreglass egg has been developed that contains electronic sensors capable of monitoring the position and temperature at six points on the shell, the humidity at the blunt pole, and the incidence of light (i.e.

whether the female is sitting or standing), and radios this information to a recorder up to 1 km away (82).

The diet of young captive birds in Hawaii contains more varied green food and less high-protein commercial poultry food than in England. Lower protein growing rations have been introduced at Slimbridge and should control the cases of 'slipped wing'* common in captive waterfowl with naturally slow growth rates.

* 'Slipped wing' is a condition shown by a few young waterfowl (of species normally breeding at low latitudes) in which the growing wing tips droop and finally twist outward. It is apparently caused by the weight of the growing flight feathers becoming too great for the strength of the muscles at the 'wrist' joint, and may be associated with 'overfeeding' (78).

5: The Release of Captive-reared Nene

By June 1978, 1,761 captive-reared Nene had been released on the islands of Hawaii and Maui (Table 2). Of these, 1,272 were set free in the native habitat on Hawaii. Of interest to us are the Nene sanctuaries on the two islands, the method of release, and the behaviour and success of the released birds.

THE SANCTUARIES ON HAWAII

The first sanctuary for the Nene was created in 1958, when a co-operative agreement was signed by the Bernice P. Bishop Estate (owners of the land), C. Brewer and Company (Lessee of the land), and the Hawaii Board of Agriculture and Forestry (now the Board of Land and Natural Resources). This Keauhou Sanctuary (Figure 36), as mentioned in Chapter 2, was the area where Elder and Woodside re-discovered the breeding grounds of the wild Nene. The agreement granted the Division of Fish and Game access to an area of 3,275 ha (8,100 acres) in order to post the area against trespass, to conduct studies on the birds, and to eliminate predators.

Most of the Keauhou Sanctuary is located at elevations between 1,825 and 2,130 m (6,000–7,000 feet) on the eastern or windward slope of Mauna Loa adjacent to the northern boundary of Hawaii Volcanoes National Park. The substrate in the sanctuary consists of *aa* (rough, angular blocks) and *pahoehoe* (smooth) lava flows of various stages of decomposition. Kipukas (see Chapter 2) are common; they vary in size and, because of their different ages, their vegetation is in different stages of ecological succession.

Lichens are the first to grow on the bare lava. Other pioneer species are ohelo, kukaenene, gosmore, pukeawe, several species of grasses and ferns, and ohia. Older kipukas, especially those in which there has been an accumulation of soil, support ohia, mamani *Sophora chrysophylla*, and koa *Acacia koa* trees, the shrubby aalii *Dodonaea viscosa*, as well as grasses and most of the plants previously mentioned. The kipukas are of special importance to the

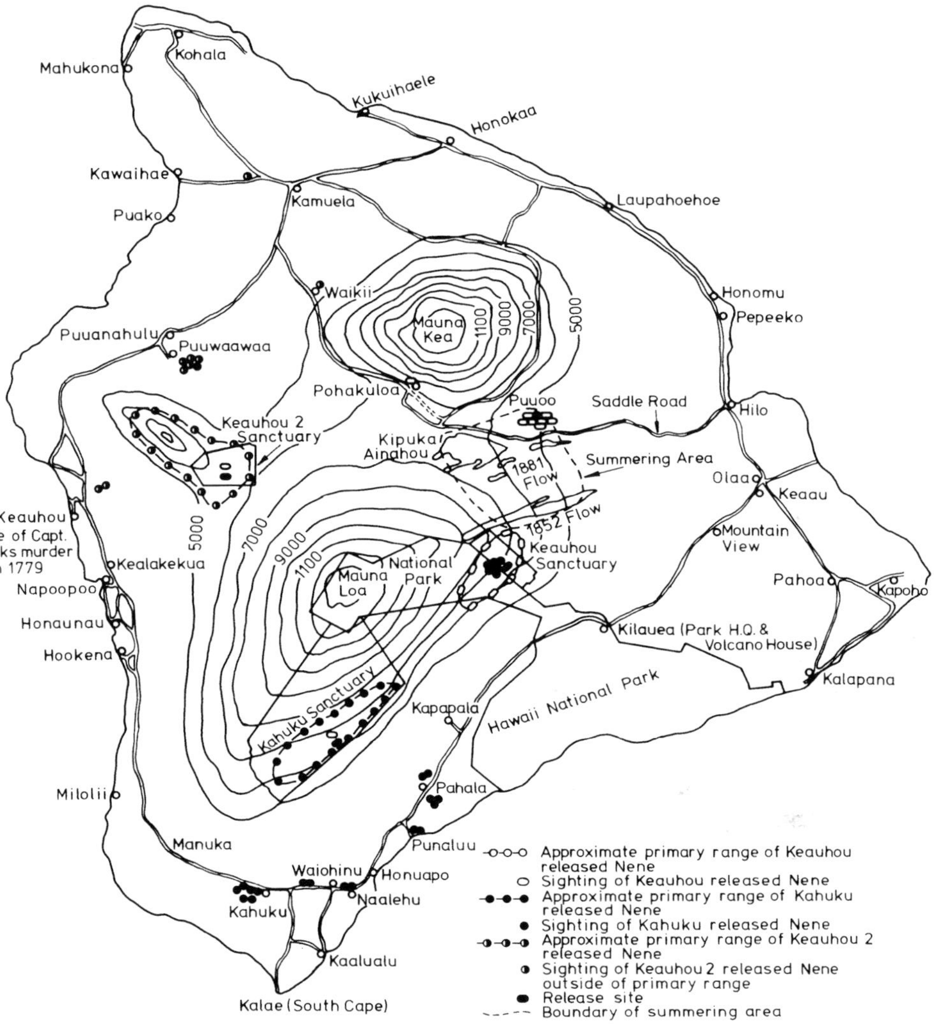

Fig. 36. Locations of the Keauhou and Kahuku Sanctuaries on the Big Island of Hawaii. Heights in feet.

Nene because they provide both food and cover.

The rate of decomposition and the rate of vegetation of lava flows in Hawaii depend largely on the annual rainfall. The Keauhou Sanctuary lies in a climatic zone characterised as one of summer drought, but the annual rainfall from 1960 to 1969 averaged 1,540 mm (60·65 inches). This pattern of rainfall favours the production of food for the Nene during the period when the birds are nesting.

Although C. Brewer and Company leases the land for cattle ranching, few cattle stray into the sanctuary itself, simply because most of it is marginal or totally unsuitable for cattle grazing. Feral goats are found, even on the sparsely vegetated lava flows, and the larger kipukas are heavily furrowed by feral pigs.

Keauhou 2 Sanctuary is at the base of Mt Hualalai and extends into the saddle region between it and Mauna Loa; it comprises 5,121 ha (12,678 acres). This sanctuary was established in 1961 by agreement with the Bishop Estate, W. H. Greenwell Ltd (the lessee), and the Board of Agriculture and Conservation. The agreement included the same provisions as those established for the first Keauhou Sanctuary.

The elevation of Keauhou 2 Sanctuary is approximately 1,825 m (6,000 feet). Here the rainfall averages about 762 mm (30 inches) per year. The habitat is noticeably different from the first Keauhou Sanctuary. Most of the area is ancient pahoehoe lava with a thin covering of volcanic cinder. The dominant vegetation consists of ohelo, pukeawe, kukaenene, aalii, and scattered small ohia trees Sandalwood *Santalum* and heau *Exocarpus* grow in the area but are not common. Kipukas consist primarily of a dry, open, mamani *Sophora chrysophylla* forest that is characteristic of this region of Hawaii. This type of kipuka is of less importance to the Nene than those found in the Keauhou Sanctuary, but the birds find food (succulent grasses and herbs) along the edges of the kipuka where fog-drip from the trees results in a good growth. Although the rainfall is low in this region, there is low cloud daily throughout much of the year.

Keauhou 2 Sanctuary is not subjected to grazing by domestic cattle, but feral goats, sheep, and pigs are sometimes found in considerable numbers. When numbers are low, they have no adverse effect on Nene habitat, but they do attract feral dogs. Moreover, feral pigs are disastrous to Nene nests, as well as to young and to the adults during their flightless stage. In January 1972, there were very few ohelo, kukaenene, and pukeawe berries. In these circumstances, the introduced wild turkeys *Meleagris gallopavo* might be significant competitors for the berries.

The Kahuku Sanctuary (Figure 36) was established in 1967 by a co-operative agreement between the Damon Estate, which owns the land, and the Department of Land and Natural Resources. This agreement permits the Division of Fish and Game to have access to 8,080 ha (20,000 acres) in order to release captive-raised birds, to conduct field studies, to eliminate predators, and to post the area against trespass.

Most of the Kahuku Sanctuary is between 1,800 and 2,150 m (6,000–7,000 feet) on the southeastern slope of Mauna Loa and above the Kau Forest Reserve. Parts of the sanctuary have a thin layer of soil which supports scattered ohia trees and an understorey of ohelo, kukaenene, gosmore, pukeawe, and several species of grasses, of which Yorkshire fog is the most common. Gosmore is abundant during the wetter months, and kukaenene is found in fruit throughout most of the year. There are some recent lava flows, with virtually no vegetation, which are used by the Nene as escape areas.

The annual rainfall at the Kahuku Sanctuary probably varies from 1,270 to 2,540 mm (50–100 inches). The nearest rain gauge (Punaluu–Kahawai) is at the upper edge of the Kau Forest Reserve, where the rainfall is presumed to be higher (Appendix 8) than in the sanctuary. Much of the precipitation within the sanctuary probably results from fog-drip from the trees.

Although the region of the Kahuku Sanctuary used to be grazed by cattle, that practice was stopped because of inadequate forage, a result of the thin soils and too little water. Feral goats are found there, but their effect on the habitat is thought to be negligible. Feral pigs, however, are abundant in the sanctuary, and so are the results of their foraging activities.

A fourth sanctuary, the Kipuka Ainahou Nene Sanctuary, was established officially during March 1974 in the area between the 1935 and 1852 lava flows, where Woodside had studied the wild Nene in 1955 (Chapter 2). This sanctuary contains approximately 15,540 ha (38,400 acres) of State-owned land and extends from the Saddle Road (running between Mauna Kea and Mauna Loa) south-easterly onto the lower flank of Mauna Loa. The elevation varies from about 1,770 m to 2,740 m (5,800–9,000 feet) and the vegetation is similar to that found in the Keauhou Sanctuary. A release pen was constructed in the sanctuary during June 1973, and 61 Nene were placed in the pen during the week of 2–6 July; 25 additional Nene were added to the pen early in 1974.

THE METHOD OF RELEASE

In his unpublished 1958 report to the Hawaii Board of Agriculture and Forestry, William Elder (47) remarked that 'with so much invested in these birds [at Pohakuloa] in time, money and hope, they must not be dumped out, without experience in flying, food finding and without knowledge of the whereabouts of the present Nene breeding ground. Experience with hand-reared waterfowl in North America has shown that a gentle release method gives much superior results to a sudden release.'

He recommended, therefore, that paired birds be confined in a large predator-proof enclosure near the Keauhou Sanctuary for several months during their normal flightless period. There they could learn to find natural foods as their mash supply was gradually reduced. As they got their full wings, they could go and return at will, until they established contact with the wild birds.

Professor Elder's general recommendations were followed, and Breese (27) described the first release of Nene into the wild at the Keauhou Sanctuary in March of 1960, when 20 birds from Pohakuloa were taken to the sanctuary. The 20 birds, 'made up of one, two, and three year olds, all raised at the Pohakuloa rearing project, were wing clipped and placed in a carefully planned release enclosure approximately one acre in size. This enclosure was built in the locality utilized by the major population of wild Nene during the nesting season, within the large Nene sanctuary. The procedure followed was to clip the primary feathers of one wing on the birds to be released and place them in the release enclosure some weeks before they were to moult. Thus, they would remain in this open topped enclosure long enough to relate to this area as well as perhaps attract wild Nene to mingle with them . . . As part of the twenty birds grew new flight feathers, a few would fly a short distance away from the enclosure, but they did not leave this area. Then, as all of the birds grew their wing feathers and could fly, the entire flock would leave the enclosure and range over this wild Nene habitat. They all remained in the vicinity of the release enclosure and did not leave the wilderness of this wild volcano slope habitat and fly to roads, towns or agricultural areas as we had feared some might. As many as ninteen of these twenty birds were identified in one day at the Nene Sanctuary in recent months.

'The question of whether they would successfully mingle with the wild Nene was also answered in the affirmative by the Nene Ecologist's recent observations late in 1960 when he observed two pairs of Nene each composed of a released bird and a wild one. He also observed a pair of Nene made up of two of the released birds, selecting a nest site near the release enclosure.'

Each of the release pens on Hawaii encloses an area between 1 and 1·5 acres (4·5 ha) (Elder had recommended enclosures of 10 acres or more.) The pens in the Keauhou (Plate 19) and Kahuku sanctuaries are built in areas where pukeawe is one of the dominant plants; associated plants include ohelo, gosmore, sheep sorrel *Rumex acetocella*, and various mountain grasses. Keauhou 2 Sanctuary is located in a much drier area where there is very little soil but a considerable amount of exposed volcanic cinder. Clumps of vegetation occur throughout the pen, and include most of the plants found in the other release pens, but there is also an abundant growth of ulei *Osteomeles anthyllidifolia*, a native Hawaiian evergreen shrub, and one or more species of the introduced mullein *Verbascum*.

Water and commercial food are provided in the pens but the amount of the latter is gradually reduced as the birds begin to eat the natural food. In order to discourage burrowing predators, the fence wire (one-inch chicken wire) is buried in an L-shape, so that the wire extends about 33 cm (1 ft) below ground level and then turns outward for another 33 cm parallel to the surface. Poison baits are placed outside the pens in an effort to save the flightless birds from such predators as mongooses, pigs, and feral dogs.

Before being released in the open-top pens, each bird is banded (ringed) with a series of coloured plastic bands. Different colour combinations are used so that every bird can be identified with certainty when seen at a later date. Wild birds are caught in the flightless state and are marked in a similar way.

PROBLEMS IN THE RELEASE

Most of the Nene are released when between two and four months of age, although some releases have contained birds more than a year old. The primary flight feathers are clipped while the immature birds are kept in the holding pens at Pohakuloa; this is done to minimize injuries that might result if birds attempted to fly in the pens. The stubs of these feathers are pulled out just before the birds are placed in the release pens. New feathers begin to grow in a short time so that most of the Nene have re-grown their flight feathers and can fly from the pen after six weeks.

In some birds, feather growth was abnormal because of damage to the feather follicles, so that gaps occurred in the flight feathers. This affected their flight ability, and some birds could not fly at all. To try to avoid this, an experiment was conducted in which the immature Nene were not wing-clipped at Pohakuloa, but their wings were brailed (the bird's 'hand' was tied to its 'forearm') before the birds were put in the release pen. Unfortunately, if brailed for only eight days, the tips of some primary feathers became twisted, and flight was impaired. Later, 24 adult Nene (that is, more than one year old) were fitted with brails for only four days; these birds sustained no damage to the flight feathers, and were able to fly when the brails were removed. For first-year birds, however, it was decided to return to the wing-plucking method, but the problem of wing damage this can cause remains. During February 1972, 20 flightless Nene were still present in the release pen in the Keauhou Sanctuary, and all had damaged primary feathers, attributed to plucking nine primaries before the birds were released during June 1971 when 110 Nene were put in the pen; 16 of these flightless birds were returned to Pohakuloa during May 1972. Similarly, of 45 Nene released in the Keauhou 2 Sanctuary on 6 September 1972, 11 flightless birds were still in the pen at the end of November, and 'even those Nene which were able to fly out of the release pen were poor flyers.' Ten Nene with damaged primaries and unable to fly were returned to Pohakuloa in December. Thirteen of the Nene with damaged flight feathers from the 1971 and 1972 releases were re-released in the Keauhou Sanctuary during May 1973; only six of these birds were able to fly from the pen a month later; the other seven birds had permanently damaged wing feathers and were taken back to Pohakuloa.

Feral dogs, pigs, mongooses, and rats *Rattus rattus* are potential predators on eggs, downy young, and flightless birds, but few actual observations of predation have been made. Dogs are seen periodically in the sanctuaries. The Nene biologist on Hawaii watched a pack of four dogs chasing a herd of sheep in the Keauhou 2 Sanctuary and on another occasion he himself was attacked by a pack of five dogs in the first Keauhou Sanctuary.

On two occasions a Hawaiian Hawk has been seen chasing an adult Nene in flight, but the outcome was not determined. Whether or not the Hawks ever prey on birds in the release pens is not known. The Hawk is an endangered species, and probably is far less common than the Nene.

SUCCESS OF THE RELEASE PROGRAMME ON HAWAII

In 1945 Paul H. Baldwin estimated that the total range of the Nene on the island of Hawaii has been reduced to about 3,000 km^2 (1,150 square miles). The present range is not known accurately but is considerably less than 2,500 km^2 (1,000 square miles). Nor do we know the size of the wild population. Published estimates are based primarily on the numbers of birds that have been released since 1960, and the figures are admittedly 'pie-in-the-sky' guesses.

There are several reasons for this lack of accurate information in a programme that is more than 30 years old. In the first place, for more than a decade no biologist with professional training in the subject of waterbirds or gamebirds has been assigned to the field work. Secondly, the habitat of the Nene, particularly on Mauna Loa, is one of the most difficult areas in the world to study. By comparison with the precipitous cliffs of Kauai, Maui, and some parts of Hawaii, the slopes of Mauna Loa are gentle, but large areas are covered by *aa* lava. Travel is possible only on foot over the sharp, angular blocks of lava, and one has to test almost every block before taking a step forward. Progress is very slow and tiring, and can be dangerous, in part because of crevices and thin-roofed lava tubes.

It is obvious that field studies by several highly competent biologists will be needed if we are ever to learn the details of the breeding cycle of the Nene in its native habitat: what is its annual production; which agents are causing mortality of eggs, downy young, and adults; and what is the size of the wild population on Hawaii, and how effective has the release programme been.

Only 17 active Nene nests were found on Hawaii during the six nesting seasons between 1965 and 1971. None of these nests was visited more than three times and 12 were visited only twice. Meaningful data on breeding biology cannot be obtained by such field methods. During the same six-year period, 31 broods containing a total of 65 young birds were observed in the field. These figures do not suggest a very successful reproductive rate in view of the 698 Nene that were released on Hawaii between 1960 and 1971.

More intensive field work is also essential because the Nene populations in the Kahuku and Keauhou 2 sanctuaries do not appear to have a daily flight pattern between feeding and roosting areas in the non-breeding season, unlike the birds from the Keauhou Sanctuary.

As described in Chapter 2, David Woodside first watched the summer flight of Nene from their feeding grounds on an 1852 lava flow to their roosting site on the Puu Oo ranch during the summer of 1955. Observations have been made along the flyway annually since that time, and as many as 114 birds (in 1969) have been counted in the Puu Oo habitat. It was not possible to check the legs of every bird for bands, but only 12 out of 80 birds closely observed carried bands

in 1969, indicating that the remainder were birds that had hatched in the wild.

Many more birds are observed on the flyway counts and on the roosting grounds than can be accounted for in the Keauhou Sanctuary during the nesting season. Birds may be coming to this summer area from as far away as South Kona. The flight paths indicate that these Nene spread out over a large area on the slopes of Mauna Loa. Many birds spend the day on the 1852 and 1881 lava flows, but efforts to locate other feeding and resting grounds have been unsuccessful.

The annual seasonal movements from the Keauhou Sanctuary usually begin in late June and continue through September, after which the birds begin to return to their nesting habitat. It is not known where the birds from the Kahuku and Keauhou 2 Sanctuaries go during this summer period.

Some of the field evidence indicates that certain Nene return to the same habitat in successive years, although there are some notable exceptions. Birds whose nests or broods are found from year to year can be censused and aged quite readily. Birds not observed at nests, however, do present problems because some have been seen for one or more years after their release, are then not seen for from two to five years, after which they may be identified again.

Homing of released birds to Pohakuloa has not been a serious problem. One goose did return to Pohakuloa twice after being released in the Keauhou Sanctuary. She was then released at the Paliku site on Maui, but again returned to Pohakuloa after a flight of about 70 miles; she was then sent to the Honolulu Zoo.

THE SANCTUARY ON MAUI

Maui is the second largest of the Hawaiian Islands; it is 42 × 77 km (26 × 48 miles) with an area of 1,880 sq km (728 square miles). Mountains at the western and eastern ends are separated by a fertile plain. The West Maui mountains are estimated to be about one million years old, whereas Haleakala, in the east end, is thought to be less than 800,000 years old. Haleakala is a dormant volcano with an elevation of 3,055 m (10,000 ft) that last erupted about 1790. Haleakala Crater is 11 km by 3 km (7 × 2 miles). Except for Koolau Gap and Kaupo Gap, the walls of the crater are about 800 m (half a mile) high.

The west end of the crater is very dry and barren, but there is an increase in rainfall and in vegetation from west to east, the east end being an area of high rainfall and rich vegetation. Here, not far from Paliku Cabin, the release pen was constructed. It covers about 2 ha (half an acre), and contains a rich growth of shrubs and herbs: for example, gosmore, sheep sorrel, mesquite grass, mountain pili *Panicum tenuifolium*. The pen itself, 1·8 m (6 ft) high, was built of 25 mm (1 inch) poultry wire, and additional wire was turned outward and buried about 30 cm (1 ft) under the ground. Before releasing any birds in the pen, the area surrounding it was heavily baited with poisoned meat in an attempt to kill mongooses and feral dogs and cats.

By 1960, enough Nene had been reared at Slimbridge (Plate 17) to make it feasible to return some of the birds to Hawaii (7; 8; 9), and the Division of Fish and Game, Department of Land and Natural Resources of the State of Hawaii, proposed that they should be used in an attempt to re-establish the species on the Island of Maui, which the Nene presumably had inhabited in the past.

'In June 1962 thirty geese were despatched by air from Slimbridge to New York. They were taken to the US Federal Quarantine Station at Clinton, New Jersey, where they were held for 21 days. They were then sent on by air freight to Honolulu. There they were kept in the Zoo for three days to recover from the journey and were given individually distinctive plastic leg bands. On the twenty-sixth day they travelled by air to Kahului, on Maui.

'The Slimbridge birds consisted of 10 juveniles, 7 one year old,

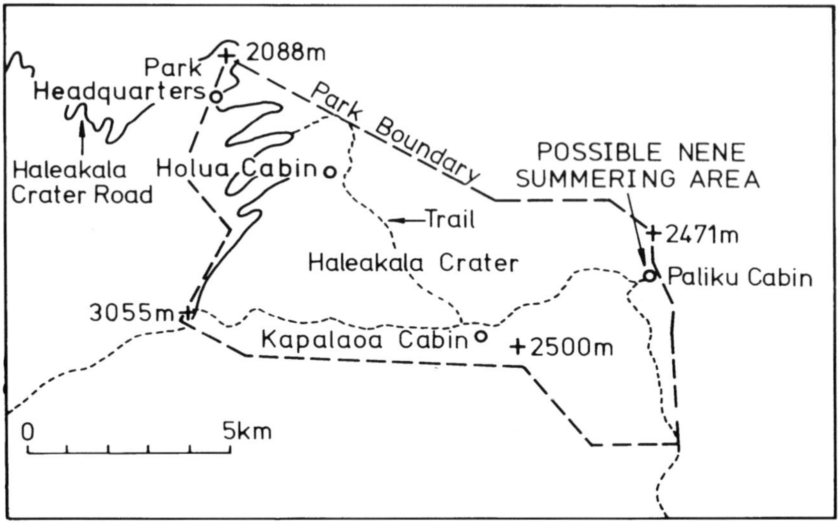

Fig. 37. Location of the release site of Slimbridge Nene in the Haleakala Crater on the island of Maui (from Ripley 110).

10 two years old, 2 three years old and 1 four years old. 17 were females and 13 males. At Kahului they were joined by five juvenile females reared by the State's Pohakuloa propagation project on Hawaii. The 35 birds were put in light-weight cardboard boxes and taken by lorry up to the rim of Haleakala Crater. Here the boxes were loaded on pack boards and were carried on the backs of 23 Boy Scout volunteers 8·5 miles down into the Crater and across to the release pen at Paliku on the other side.

'The release pen at Paliku encloses about an acre of good grass* on the edge of an ancient "aa" lava flow. The geese were released in the late afternoon of July 26. As they were removed from the boxes, their clipped primaries were extracted so that new feathers would start to grow immediately. The birds were given constant care and supplied with native berries and scratch feed daily. They were also treated for coccidiosis and caecal worms – these parasites having been detected in the faecal droppings of the flock. There was much pecking and other signs of social adjustment for a week or so, but the birds soon settled down.

'The geese began making short flights within the pen early in

* The pen was actually nearer ½ acre (2,000 m^2) in size.

September. The first flights out of the pen were seen on September 12, two Hawaiian-reared birds being the pioneers. The first Slimbridge bird did not fly out until September 17. By the end of the month, 27 out of the 35 were flying a few hundred yards out of the pen, returning at dusk. The Hawaiian and English birds, though remaining in distinct flocks within the pen, mixed very well outside it.'

Other releases were made during the following years, and 489 Nene had been given their freedom in Haleakala Crater by July 1978 (Table 2). Because the early Slimbridge birds were the progeny of the Shipman-strain geese, it seemed advisable also to release some birds from Pohakuloa that had been derived, in part, from the wild-strain stock. Thus far, 197 birds from Slimbridge (3 geese died before release), 285 from Pohakuloa, and seven from the collection of S. Dillon Ripley in Connecticut, have been freed in the crater. Most of the releases have been made during the summer months (late June to August). Carriage on the back of Boy Scouts or of personnel of the State Division of Fish and Game has been replaced by a mule pack-train and, during 1972, by a helicopter.

BEHAVIOUR AND SUCCESS ON MAUI

Although the first group of 35 Nene was placed in the release pen in Haleakala crater during the summer of 1962, they did not leave when their wing feathers grew out: most remained in the area until April 1963. A gradual departure was recorded after that, but six birds remained until late June, and two birds still occupied the habitat when the second release was made.

Beginning with the release in 1963, the behaviour of the Slimbridge Nene was more typical than in the first release of behaviour found among the birds on Hawaii. Placed in the pen during the summer, some began to leave the release area by the end of November, and all had left by the end of December. Moreover, by 5 September 1963, nine of the birds that had been released in 1962, and that had not been seen for several months, returned to the Paliku region and mingled with the younger birds. Again in 1964, the birds began to disperse from the release site by mid-November.

The first nests were not found in Haleakala Crater until 1968. Between then and 1971, 38 nests were found in the general area of the release pen. Only 11 young birds were known to have hatched from these 38 nests, and five of these were found dead. Apparently only three of the young survived to the flying stage. These were seen with their parents (a Slimbridge and a Pohakuloa bird) during May 1971 (106); one of the young was paired with another Slimbridge bird in January 1972, and a second bird was seen in 1973. No nests were found during the 1971/1972 breeding season, however. Fifty-one Nene were seen in the vicinity of the release area during April 1972, but there was no indication of successful reproduction.

It is possible, therefore, that between 1962 and January 1972, no more than two young Nene had survived long enough to enter the breeding population in Haleakala. However, State Fish and Game personnel believe that eight young may have survived to adulthood during the 1972/1973 breeding season. Only two goslings were raised to maturity during the 1977/1978 breeding season.

It should be pointed out that the 8,605 ha (21,300 acres) in Haleakala Crater that are thought to be suitable habitat consist of exceedingly rough terrain (53). The full-time effort of one or more qualified biologists will be required to obtain an accurate census of the number of surviving Nene and to discover the factors that are contributing to the failure of the reproductive cycle. The State Division of Fish and Game, however, has never assigned a full-time biologist to such a task.

Violent winds, rain, and several hail storms caused mortality during some years. For example 65 cm (25·5 inches) of rain fell between 4 and 10 January 1969, and 91 cm (36 inches) of rain, accompanied by hail, were recorded for the three-day period ending on 1 February that year. All but two of 15 nests were abandoned during this period. Three goslings hatched on 28 January 1969, but all were dead a week later; these were the first goslings ever seen on Maui since the first Nene were released there in 1962. Two eggs hatched in one other nest, but the young did not survive. By contrast, severe droughts occur in some years.

Although snow is infrequent on Haleakala, some of the heaviest snowfalls recorded during the past 40 years have occurred since 1962. The role of predators in the destruction of eggs, downy young, and flightless birds has not been determined.

THE NATIONAL PARK PROGRAMME

The National Park Service and the State Division of Fish and Game initiated a co-operative programme in 1972 to attempt to re-establish the Nene in its former range within Volcanoes National Park and to establish a small population in areas more accessible to the public in Haleakala National Park. Sir Peter Scott (108; 118) had proposed the creation of 'Nene Parks' on a much grander scale: 'For each Park an area of about 30 acres would be required which must be surrounded with a cat-dog-and-pig-proof fence to a height of 8 feet. The possibility of making it mongoose-proof should also be considered.' However, the cost of constructing such a pen (with '10 to 15 small enclosures of about half an acre apiece, each to accommodate a breeding pair of Nene') was prohibitive, and there is no known way to build a mongoose-proof, open-topped pen.

Nine open-topped pens from 4,000 m^2 to 8,000 m^2 in size (one to two acres) had been constructed at Volcanoes National Park by 1 November 1978. A breeding pair, made flightless by pinioning or by clipping primary feathers, was maintained in each enclosure. Goslings were permitted to fly free so that they could leave or return at will. Although mortality of goslings had been high, by the summer of 1979 at least 30 Nene had survived to adulthood in the Park.

Two small pens were built adjacent to the Haleakala National Park headquarters building at 2,150 m (7,000 ft) on Maui; a pair of Nene was placed in each pen. A release pen 30 × 30 m (100 × 100 ft) also was built on the slopes of Haleakala between Hosmer's Grove and Park Headquarters. Although the caged Nene were viewed daily by visitors to the Park, one pair hatched five goslings during December 1972. They were placed in the release pen as soon as they became independent, and have since flown from it. Four goslings were produced by the two pairs during the 1973/1974 breeding season. A third pen was then built. However, only one gosling was raised from 17 eggs laid during the 1977/1978 season.

P.S.
1957

6: The Experiment in Conservation

The example of the Nene (along with that of Père David's Deer *Elaphurus davidianus*, Przewalski's Horse *Equus przewalskii* and the European Bison *Bison bonasus*) is often quoted as one of the few instances where a species has been saved from extinction by captive breeding. The propagation programme, which has resulted in the restocking of the wild, has become something of a conservation legend. But has it been an unqualified success?

We may state with confidence that there are far more Nene alive now than there were a quarter of a century ago. In the summer of 1973, the Wildfowl Trust held, or had on loan to other zoos and private breeders, 432 birds. At that time it was decided to ask the US Department of the Interior whether further 'English' birds were needed for release in Hawaii and, if not, whether surplus stock could be sold. This latter was agreed to, and Nene are now available to all aviculturalists; nearly 130 pairs have been sold, and many of these are now breeding elsewhere. Numerous other Nene in North American zoos and collections originate from a captive breeding flock maintained at Patuxent, Maryland, by the US Department of the Interior's Rare and Endangered Species Unit (1; 59). Over 50 birds are in captivity in Hawaii and the annual production at Pohakuloa is now about 50 (Table 2). There are, at a guess, 750 birds in the wild and 1,250 in captivity. The Red Data Book of the IUCN now includes the Nene in the list of Rare, rather than Endangered, birds. Thus, barring catastrophe, the species has been saved from extinction. In that aim, the programme has obviously succeeded, and succeeded extremely well.

Nevertheless, we are justified in asking whether the thousand and a half released birds have increased the breeding potential of the wild Hawaiian Goose, of which so few remained in 1949. Is the native population now better able to sustain itself indefinitely? It seems likely that it is, but as the last chapter will have indicated, we do not really know the answer. The basic recommendation of Smith (120) and Elder (47; 48) for an intensive field study has yet to be followed. We still do not know, other than in general terms, what brought the species so low, and so cannot be sure that the hazards have been removed or are being effectively controlled. If *nothing* had been done in 1949, it seems likely that the Nene would be extinct. But what if the money used to produce birds for release had been used in predator control instead? If the pig and goat clearance, and poisoning of dogs and mongooses, had started earlier on a large scale, might the wild population have recovered without additions? Even before releases were made, numbers in the wild seemed to be rising a little – or at least human estimates of the size of the population were. In 1950, only about 20 wild birds were thought to exist (6; 7). Breese in 1957 (26) pointed out that there were large areas of range in which almost no observations had been made and suggested that the previous estimate of 30 (7; 120) was too low. Elder in 1958 (48) was putting the number at 'at least 35', and by 1960, before releases started, there were thought to be either 49 or 50 birds (71; 117). It is probably fruitless to speculate – the release of captive birds *was* initiated, and adequate field studies are still required. Because of this inadequacy of information, it is impossible to be definite about the outcome of the reintroduction phase of the programme. We are still groping for answers, and can only say 'may be', 'probably' and 'perhaps'. This chapter is full of such qualifications, but tentative conclusions are made, and we hope that they will be useful.

At least we now know a great deal more about captive breeding programmes. Some of the problems that will be encountered were admirably covered in a paper (95) by Professor G. V. T. Matthews, Director of the International Waterfowl Research Bureau, and the following paragraphs will examine a few of these using the Nene as an example.

The general observation is that small captive stocks deteriorate, especially in terms of fertility, when it is not possible to bring in a supply of new genetic material from a wild population (95). This seems to have happened last century to the captive Nene population in Europe (Chapter 3) and to the 'Shipman' stock more recently. In the case of the Slimbridge birds, four new males have been introduced; two came after the programme had been going 12 years (in 1962) and two more in 1966.

In restricted, inbred populations, genes for deleterious characters, which are generally recessive, are more likely to be brought together and their effects emerge in the offspring. An example is the thin-down condition in the Nene gosling (Plate 23). At Slimbridge, a

quarter of the offspring of Emma and Kamehameha showed the trait (Figure 25) but, with adequate care, survived to reproduce themselves. Both Emma and Kamehameha must have carried the recessive gene – not surprising, perhaps, as the birds were closely related and came from the original 'Shipman' stock (Plate 1). However, 'cottony' down also appeared in the descendants of one of the males of wild stock (from the pair captured in March 1960 – Chapter 3) that had been sent from Hawaii to Slimbridge in 1962. He was a particularly fertile gander, and thus he and his offspring figured largely in the breeding programme, with the result that the down condition became quite common and by 1973 was occurring in 20% of all goslings. The gene has occurred also in the wild stock held at Pohakuloa, where any pairs that produce the deficient goslings are eliminated 'unless a parent is from our pure wild strain. In that case the wild bird is mated to another bird' (D. H. Woodside, in litt.). The wild bird must also be a carrier, because only if both parents have a recessive gene can the character appear in their offspring. This illustrates one of the problems of captive breeding programmes – that positive selection for one feature (fertility or 'wildness') may increase another less desirable recessive trait (young with meagre down). Perhaps thin-down was no disadvantage in the original Nene population that descended to lower altitudes to breed; the climate may have been mild enough for affected young of this non-swimming species to survive; indeed, it is conceivable that thin-down was useful. Brambell (25) has suggested that this type of natural variation should be preserved in captive rearing programmes: 'presumably the original population contained downy and less downy genes, giving it greater flexibility in colonising cold and wet places as well as hot and dry ones'. However, one wonders whether thin-down is not now implicated in the species' lack of success in the wild: habitat destruction, the presence of the mongoose and perhaps of the mosquito, have forced the birds to breed at colder, wetter altitudes, and at least one 'cottony' gosling has recently been found dead in the wild (143). The gene has certainly been augmented in this upland-breeding population, as early releases on both islands were almost entirely of 'Shipman' stock. Personnel of the State Division of Fish and Game reported that three birds that were 'cottony' when hatched were unable to adapt to the release pens and had to be returned to Pohakuloa in 'an emaciated condition'. As already noted, efforts have been made at Pohakuloa to eliminate the carriers of the condition. Lee reported that affected goslings there are 'weak and sickly' (143), and those at the London Zoo left in outdoor pens with their parents have not survived (P. J. Olney pers. comm.). At Slimbridge, positive elimination has not been thought necessary as most of the birds are destined for captivity.

Many natural selection pressures, as well as those imposed by inclement weather, are removed during captive breeding. Predators and food shortage are seldom met with. An aviculturalist is justly proud if he raises a complete brood from hatching, but he will in fact be dulling the edge of natural selection which normally removes, say, 70% of the least effective young (95). Few aviculturalists are sufficiently ruthless to eliminate those showing defects if a species is very rare. Few are even sufficiently methodical to avoid breeding from such birds. In the thin-downed Nene, any elimination is rendered more difficult because the character is not obvious in the adult.

Not only can birds carrying deleterious genes survive in captivity, but there may be inadvertent selection for characters that stand the captive animal in good stead and yet would be disadvantageous in the wild. Only those individuals that tolerate captivity will breed, and they may do so because they have a lower reaction to sudden noises and strange objects than their fellows. They are clearly 'tamer', in that they will allow human proximity during the breeding season. It follows that individuals that cannot settle in captivity seldom reproduce, and therefore tend not to pass on their 'wild' genes. For this reason propagation should be under conditions as natural as possible. Emphasis should be laid on the production of sound stock rather than on the rapid build up of numbers. There is evidence that keeping Red Grouse *Lagopus lagopus* on pelleted rations for many generations will produce a line of birds with naturally short guts, since those with long guts do not live as long (81). Changes in territorial behaviour can also be selected for in captive animals maintained in relatively small enclosures, with abundant food (95). Yet released into the wild, the acquisition of a territory may be essential, and the ability to digest the natural diet seems equally vital.

It is also preferable for the propagation unit to be as close to the original habitat as possible. The further away it is, the more likely are novel selection pressures to be brought to bear. Thus, if birds from low or high latitudes are bred in more temperate zones (where most of the 'good' zoos are situated), selection will favour those individuals that survive and breed under a different range of temperatures, rainfalls, and daylengths (81; 95). If requirements at the home site call for a particularly precise timing of the breeding season, such flexibility may be very unhelpful on release. As has been shown in the Nene, growth rates of the young can vary at different latitudes, and initial growth rate *may* be related to the ultimate life span. Clearly Slimbridge's great expertise with waterfowl would have been put to even better use nearer the original home of the Hawaiian Goose where the breeding season is so much longer.

Again, certain birds, especially those from remote islands, and, therefore, often in the Rare or Endangered categories, show poor resistance to infection in zoos in North America and Europe. If sufficient individuals are taken into captivity, enough may survive to set up a disease-resistant stock in due course. But this population could have a very different genetic composition from the original one, and might be ill-suited for conditions encountered on release in the native habitat. The rare Rothschild's Starling *Leucopsar rothschildi*, which in the wild is restricted to the island of Bali, has been found to be unusually susceptible to avian pox transmitted by common Starlings *Sturnus vulgaris* (81). Only when birds at the National Zoo in Washington DC were isolated did mortality cease. The role of introduced diseases in the extinction of the endemic Hawaiian avifauna is considered by some to have been an important one (130), although there is little reliable evidence to support this 'logical conclusion' (18). However, a similar situation to that observed at Washington Zoo arose when certain native Hawaiian birds were kept at Honolulu Zoo (130). Pox was probably absent from Hawaii until the introduction of numerous songbirds that carried the infective organism to the islands in their blood. The infecting agent is thought to be the mosquito *Culex pipiens fatigans*, introduced accidentally into Maui in 1826 from a ship cleaning out its fresh-water tanks. The mosquito spread and is present today on all the Hawaiian islands, but is most common in areas below 300 m (1,000 ft). Fortunately, Pohakuloa is situated at 1,980 mm (6,500 ft); however, at the Wildfowl Trust in England, a dozen Nene (alone among the waterfowl kept there) have suffered from a serious condition that appears to be avian pox (83). Perhaps the English mosquito is spreading the disease to them (Plate 23). There is, of course, the additional risk that captive-reared birds are carriers of disease organisms that will be disastrous to the remaining wild stock upon their reintroduction. It was partly to guard against this possibility that the Nene from England were returned to Maui rather than to the island where the last wild birds remained.

Leaving aside the long-term effects of captivity on the genetic make-up and pathological condition of the propagated species, we should also consider its immediate effects on the behaviour of the individual. In particular, waterfowl have that peculiar form of learning, known as 'imprinting'. This involves a bird learning the characteristics of its parents shortly after hatching. Many aviculturalists use domestic hens to rear their stock, or raise their birds by hand. But human-reared animals are often quite unsuitable for liberation into an environment where Man is still a predator. Being sociable, geese learn much of their behaviour from older birds. If older generations are lacking, either during the rearing stage or at the release point, then the young birds may wander and fail to develop the necessary behaviour patterns of flocking and dispersal. Certainly, it seems that success in the release of captive-reared geese is more likely with a near-sedentary population such as the Nene, and when some wild birds are already established at the site, such as on the Big Island of Hawaii. Also, it is probably important to release birds as soon as they are physically able to fend for themselves. The early Nene releases contained a proportion of birds that had been in captivity for more than one year. These older birds may be less able to adjust to novel circumstances, and it is significant that the first known successful breeding on Maui was of a pair (one from

Pohakuloa and one from Slimbridge) released when only a few months old.

When the objective is the restoration of a population to its original habitat and locality, the habitat ought to be in a suitable state to receive the re-introduction. It is little use releasing large numbers of a rare waterfowl species if the cause that led to its past extinction is still present, or some other adverse factor has developed in the interval. If, for instance, introduced predators such as the pig and the mongoose were limiting factors, the predators need to be removed or at least kept to a low level. And, if it is not possible to remove the predators completely, control of their numbers has to continue *indefinitely*. These, and many other provisos, imply that the agency that is attempting the captive breeding programme should have effective control of the habitat to which the stock is to be restored: one more argument for having the whole project set up in the country of origin wherever practicable. In the case of the Nene, it was not easy to put the bird back into the habitat where the majority of the population once bred. The lowlands are now largely unavailable to it, and it seems that the highlands, especially of Maui, may not supply an optimal breeding environment. Given that the habitat *does* provide a secure base for eventual release, a captive breeding programme involving waterfowl has a good chance of ultimate success, since they often take readily to captivity and breed freely. There are, indeed, at least nine captive breeding programmes for re-introduction into the wild that recently or currently involve waterfowl species (79). These include the Hawaiian Duck or Koloa, which was almost eliminated by the introduced predators, and by hunting. A breeding programme was started at Pohakuloa, and a number of birds have been released on Hawaii and Oahu.

How should the restoration be done? The idea of using a gradual settlement technique was put forward by Sir Peter Scott (118) for the Nene, and fifteen years later the plan is being implemented in modified, rather small-scale form as described in Chapter 5. Really substantial enclosures are needed and a considerable amount of time and money has to be spent by the propagating authority in the construction and running of a waterfowl park. Because admittance to US National Parks is free, the expense of a Nene Park could not be offset by revenues from tourists, as might be the case in other areas. This was probably the reason why even the modified plan took so long to be realised. However, Nene can now easily be seen at Pohakuloa and Volcanoes National Park on Hawaii, and at Haleakala National Park Headquarters on Maui. Enclosures have also been set up with breeding pairs in remote locations at lower elevations not now used by Nene; the goslings will be left free-winged, and perhaps will inhabit a small part of the former Nene range (W. E. Banko, pers. com.). Gradual resettlement may be shown eventually to be preferable to any other restoration technique. The habitat occupied by the remnants of the wild Nene population at the moment is too remote and rugged to be accessible to any but the most dedicated bird-watching tourists.

Finally, any re-introduction programme needs an adequate series of follow-up surveys. These ought to indicate that birds are surviving in reasonable numbers and are beginning to breed successfully under wild conditions. Otherwise the programme is failing in its objective. If a shortage of funds does not permit simultaneous running of the propagation scheme and wild surveys, then the former should, perhaps, be interrupted until the latter have been carried out. It is no use continuing to pour expensively propagated birds into the wild if the reintroduction is not 'taking'. Unfortunately, surveys of the Nene population have been inadequate, partly because the terrain is so difficult. Of 340 birds released on to Maui only one pair was *known* to be breeding successfully by January 1972. This is not to say that other birds are not now reproducing (53; 92), but it is very difficult to be certain of the extent of success or failure. On the Big Island of Hawaii where some wild stocks remained, massive releases of hundreds of birds reared nearby appear to be producing more positive results, but again really detailed information is lacking.

If we are to agree, with hindsight, that Nene for release would have been better propagated at the centre deliberately set up for that purpose at Pohakuloa (or better still, as Sir Peter Scott suggested, at a number of really large Nene parks on Hawaii or Maui), what then is the role of zoos in the saving of endangered species? Without the scientific programme carried out at Slimbridge over the last 30 years, the knowledge that breeding projects at another latitude and in a different climate could produce difficulties would never have emerged. The role of zoos, perhaps, is three-fold; first, they should give practical help and advice, as indeed happened repeatedly in the case of the Nene, and secondly carry out and encourage research into basic biology and husbandry. More knowledge is required about the biology of most rare and endangered animals. In certain cases, this can be better obtained from captive individuals, and later applied to their undisturbed wild relatives. It is as important to know what *not* to do as it is to know what to expect. Here zoos can come into their own, although they must keep careful records over many years if information of value is to emerge. Only after these basic data have been collected and examined can we know what an animal species is capable of in terms of territoriality, reproductive potential, and so forth. Along with our wish to save wild creatures in their natural state should go a desire to improve the husbandry of all animals. Half a century ago it was probable that many zoo inhabitants were undernourished; now comes the distinct possibility that others (with the best of intentions) are being over-fed, their life expectancy shortened and their health undermined by diets that are too 'rich'. This knowledge comes after much record-keeping, experience, and research. Of course, we must make a distinction between captive stock studied scientifically and 'zoos' in the usually accepted sense of the word – the exhibition of captive animals for public recreation. This third major role of the zoo is to introduce the mass of the people, who will never go into wild places, to the animals that their money, in the form of taxes or donations, will increasingly be required to support. One direct achievement of the project to save the Nene has been the arousal of tremendous public interest and, thereby, an increased awareness of the value and possibilities of conservation programmes in general. It was partly for this reason that pairs of Nene have been widely distributed to zoological gardens and wildlife collections in many parts of the world (37; 57; 59; 86; 90). As well as making attractive and educational exhibits, the birds help to ensure that no single disaster overtakes the whole population in the future.

Conservation is generally expensive and it is not always easy to decide how to allocate funds that are available. Initially, money should perhaps be used to purchase reserves in which rare animals can continue to live undisturbed. These reserves may need to be vigorously protected, both from Man and from his domestic and introduced animals and plants. Sometimes, it may be advisable to supply artificial food to the endangered animal at times of shortage, or to construct artificial nesting sites where they are lacking. In the case of birds, eggs may be collected, and the young raised and released at an early age into the habitat. More animals can survive to adulthood by this method than if the parents rear their own young, but it is probably only sound policy for a species with relatively little dependence on traditional learning. Only as a last resort, as with the Nene, should animals be taken into captivity, bred and released.

Appendix 1: Endemic, extinct, rare and endangered Hawaiian birds

A: ENDEMIC HAWAIIAN BIRDS

PROCELLARIIDAE (shearwaters, petrels, and fulmars)
Newell's Manx Shearwater *Puffinus puffinus newelli*
Dark-rumped Petrel *Pterodroma phaeopygia sandwichensis*

HYDROBATIDAE (storm petrels)
Harcourt's Storm Petrel *Oceanodroma castro cryptoleucura*

ANATIDAE (ducks, geese, and swans)
Nene or Hawaiian Goose *Branta sandvicensis*
Koloa or Hawaiian Duck *Anas platyrhynchos wyvilliana*
Laysan Duck, *Anas p. laysanensis*

ACCIPITRIDAE (hawks, kites, and eagles)
Io or Hawaiian Hawk *Buteo solitarius*

RALLIDAE (rails, crakes, gallinules, and coots)
Laysan Rail *Porzanula palmeri*
Moho or Hawaiian Rail *Pennula sandwichensis*
Alae or Hawaiian Gallinule *Gallinula chloropus sandvicensis*
Alaekea or Hawaiian Coot *Fulica americana alai*

RECURVIROSTRIDAE (avocets and stilts)
Aeo or Hawaiian Black-necked Stilt *Himantopus h. knudseni*

STRIGIDAE (typical owls)
Pueo or Hawaiian Short-eared Owl *Asio flammeus sandwichensis*

CORVIDAE (crows, jackdaws, jays, and magpies)
Alala or Hawaiian Crow *Corvus tropicus*

TURDIDAE (thrushes, robins, wheatears, and allies)
Puaiohi or Small Kauai Thrush, *Phaeornis palmeri*
Omao or Hawaiian Thrush, *Phaeornis obscurus* (5 subspecies)

SYLVIIDAE (Old World warblers)
Laysan Millerbird, *Acrocephalus f. familiaris*
Nihoa Millerbird, *Acrocephalus f. kingi.*

MUSCICAPIDAE (Old World flycatchers)
Elepaio, *Chasiempis sandwichensis* (3 subspecies)

MELIPHAGIDAE (Old World honeyeaters)
Kauai Oo, *Moho braccatus*
Oahu Oo, *Moho apicalis*
Molokai, Oo, *Moho bishopi*
Hawaii Oo, *Moho nobilis*
Kioea, *Chaetoptila angustipluma*

DREPANIDIDAE (Hawaiian Honeycreepers)
Amakihi, *Loxops virens* (4 subspecies)
Anianiau, *Loxops parva*
Greater Amakihi, *Loxops sagittirostris*
Creeper, *Loxops maculata* (6 subspecies)
Akepa, *Loxops coccinea* (4 subspecies)
Kauai Akialoa, *Hemignathus procerus*
Akialoa, *Hemignathus obscurus* (3 subspecies)
Nukupuu, *Hemignathus lucidus* (3 subspecies)
Akiapolaau, *Hemignathus wilsoni*
Maui Parrotbill, *Pseudonestor xanthophrys*
Ou, *Psittirostra psittacea*
Laysan Finch, *Psittirostra c. cantans*
Nihoa Finch, *Psittirostra c. ultima*
Palila, *Psittirostra bailleui*
Greater Koa Finch, *Psittitrostra palmeri*
Lesser Koa Finch, *Psittirostra flaviceps*
Grosbeak Finch, *Psittirostra kona*
Apapane, *Himatione s. sanguinea*
Laysan Honeycreeper, *Himatione s. freethii*
Crested Honeycreeper, *Palmeria dolei*
Ula-ai-hawane, *Ciridops anna*
Iiwi, *Vestiaria coccinea*
Mamo, *Drepanis pacifica*
Black Mamo, *Drepanis funerea*
Black-faced Honeycreeper, *Melamprosops phaeosoma*

B: EXTINCT HAWAIIAN BIRDS (18)

Laysan Rail *Porzanula palmeri*	
Hawaiian Rail *Pennula sandwichensis*	
Oahu Oo *Moho apicalis*	
Molokai Oo *Moho bishopi*	
Hawaii Oo *Moho nobilis*	
Black Mamo *Drepanis funerea*	Molokai
Kioea *Chaetoptila angustipluma*	Hawaii
Greater Amakihi *Loxops sagittirostris*	Hawaii
Greater Koa Finch *Psittirostra palmeri*	Hawaii
Lesser Koa Finch *Psittirostra flaviceps*	Hawaii
Grosbeak Finch *Psittirostra kona*	Hawaii
Ula-Ai-Hawane *Ciridops anna*	Hawaii
Mamo *Drepanis pacifica*	Hawaii
Akialoa *Hemignathus obscurus*	(all three subspecies extinct on Oahu, Lanai and Hawaii)
Laysan Millerbird *Acrocephalus f. familiaris*	
Laysan Honeycreeper *Himatione sanguinea freethii*	
Oahu Thrush *Phaeornis obscurus oahensis*	
Oahu Akepa *Loxops coccinea rufa*	
Oahu Nukupuu *Hemignathus l. lucidus*	
Lanai Thrush *Phaeornis obscurus lanaiensis*	
Lanai Creeper *Loxops maculata montana*	
Extinct populations of surviving species	
Iiwi *Vestiaria coccinea*	Lanai
Ou *Psittirostra psittacea*	Oahu, Molokai and Lanai
Crested Honeycreeper *Palmeria dolei*	Molokai

C: RARE AND ENDANGERED HAWAIIAN LAND BIRDS (18)

Hawaiian Hawk *Buteo solitarius*
Hawaiian Crow *Corvus tropicus*
Small Kauai Thrush *Phaeornis palmeri*
Large Kauai Thrush *Phaeornis obscurus myadestina*
Molokai Thrush *Phaeornis o. rutha*

Nihoa Millerbird *Acrocephalus familiaris kingi*
Kauai Oo *Moho braccatus*
Kauai Nukupuu *Hemignathus lucidus hanepepe*
Kauai Akialoa *Hemignathus procerus*
Kauai, Maui and Hawaii Ou *Psittirostra psittacea*
Oahu Creeper *Loxops m. maculata*
Oahu and Molokai Iiwi *Vestiaria coccinea*
Molokai Creeper *Loxops maculata flammea*
Lanai Apapane *Himatione s. sanguinea*
Lanai Amakihi *Loxops virens wilsoni*
Maui Akepa *Loxops coccinea ochracea*
Maui Black-faced Honeycreeper *Melamprosops phaeosoma*
Maui Nukupuu *Hemignathus lucidus affinis*
Maui Crested Honeycreeper *Palmeria dolei*
Maui Parrotbill *Pseudonestor xanthophrys*
Hawaiian Creeper *Loxops maculata mana*
Hawaiian Akepa *Loxops c. coccinea*
Akiapolaau *Hemignathus wilsoni*
Palila *Psittirostra bailleui*
Laysan Finch *Psittirostra c. cantans*
Nihoa Finch *Psittirostra c. ultima*

Laysan Duck *Anas p. laysanensis*

Appendix 2: Weight of adult Nene (g)

AT SLIMBRIDGE

♀ as % of ♂	Month	MALES Mean weight	No. in sample	Standard error(±)	Signif.	FEMALES Mean weight	No. in sample	Standard error(±)	Signif.
94	January	2,070	47	22·40	NS	1,945	41	26·65	–
–	February	2,045	8	71·68	NS	–	0	–	–
–	March	2,050	8	44·58	*	1,935	2	144·99	NS
88	April	2,335	39	49·88	NS	2,055	28	47·81	NS
85	May	2,295	33	37·89	‡	1,945	21	46·00	†
88	June	2,060	61	21·61	NS	1,820	41	24·81	NS
84	July	2,115	31	28·51	NS	1,785	25	30·99	‡
86	August	2,195	73	26·39	NS	1,895	69	22·36	NS
86	September	2,155	47	36·13	NS	1,935	36	36·60	NS
84	October	2,230	51	29·45	NS	1,960	41	29·43	NS
86	November	2,240	57	28·15	NS	2,005	49	27·57	NS
88	December	2,195	35	31·18	†	1,930	28	40·20	NS
	January	2,070	47	22·40		1,945	41	26·65	

The 'mean' Slimbridge male weighs 2,165 g, and the extremes range from 1,695 g (in June) to 3,050 g (in April). The 'mean' Slimbridge female weighs 1,930 g and the extremes range from 1,525 g (in July) to 2,560 g (in April).

KEY to Standard error.
NS means weight not significantly different between one month and the next.
* ($P < 0.05$) means weight significantly different, but only slightly so.
† ($P < 0.01$)
‡ ($P < 0.001$) means weight very different.

AT POHAKULOA (10):

	MALES Mean	No. in sample	FEMALES Mean	No. in sample
'After moulting'	1,985	26	1,800	26
October	2,010	26	1,925	26

Males range from 1,675 to 2,270, and females from 1,575 to 2,240.

Appendix 3: Average linear measurements of Slimbridge adults (mm)

	Sex	*Mean*	*No. in sample*	*Range*	*Standard error*
Wing	♂	378	30	351–404	2·36
	♀	361	30	347–368	1·88
Bill	♂	39·4	30	36·0–43·1	0·31
(Culmen)	♀	37·0	44	31·9–40·0	0·31
Head	♂	93·5	31	90·2–98·8	0·43
(Bill tip to back of skull)	♀	89·2	30	84·8–94·5	0·44
Tarsus length	♂	85·0	31	76·0–89·6	0·45
	♀	78·0	31	73·8–82·9	0·43
Tarsus thickness	♂	7·43	62	7·0–8·1	0·04
	♀	7·06	62	6·5–7·9	0·04

The recommended ring or band size for Nene is 16 mm (internal diameter).

Appendix 4: The egg of the Nene

SLIMBRIDGE FRESH WEIGHTS (g)

Description	*No. in sample*	*Mean*	*Range*	*Standard error*	*Signif.*
All eggs	700	144·0	(88·0–173·0)	0·43	‡
1st egg of clutch	194	141·0	(99·5–171·0)	0·85	
Eggs of 1st clutch	403	145·1	(88·0–172·0)	1·61	NS
Eggs of 2nd clutch	137	143·3	(99·5–173·0)	0·94	NS
Eggs of 3rd clutch	9	141·3	(125·0–152·0)	2·59	
Infertile eggs	133	141·3	(95·0–163·5)	0·99	‡
Eggs that hatched	187	146·6	(116·0–168·0)	0·67	
Eggs laid by					
2 year olds	90	144·0	(99·5–164·0)	1·11	NS
3 year olds	145	146·1	(121·0–173·.0)	0·83	NS
4	112	146·6	(104·5–166·5)	0·89	†
5 and 6	174	143·1	(95·0–164·0)	0·09	‡
7 and 8	61	138·1	(88·0–157·0)	1·43	NS
9 and over	37	139·7	(106·5–170·5)	2·45	

POHAKULOA FRESH WEIGHTS (g) (10)

Eggs of 1st clutch	128	150·7	(120·9–181·5)	–	–
1st egg of clutch	31	149·6	–	–	–
Infertile eggs	11	146·7	–	–	–
Eggs that hatched	88	151·4	–	–	–
Calculated fresh weight (g) (113) From linear measurements of eggs probably laid in captivity in Europe	10	131.0	–	–	–

EGG DIMENSIONS (mm)

Mean	*Range*	*No. in sample*	*Source*
78·2 × 55·0	(74–89 × 52–58)	10	Captive in Europe? (113)
80 × 50	–	–	Captive in Europe? (39)
83·1 × 55·6	(68·2–98·1 × 49·0–64·3)	300	Captive at Slimbridge
84·0 × 62·0	–	–	Wild (36)
85·4 × 59·7	–	3	Wild (62)

SHELL WEIGHTS (g)

Mean	*Range*	*SE*	*No. in sample*	*Source*
10·6	(9·0–11·9)	–	10	Captive in Europe? (113)
12·25	(8·21–15·07)	0·12	125	Captive at Slimbridge
13·0	(12·7–14·2)	–	6	Captive at Slimbridge (C. Tyler pers. com.)
16·4	(12·1–23·9)*	–	87	Captive in Hawaii (10)

* Shells weighed after hatching; all other weights are of blown eggs

SHELL THICKNESS (mm)

No. in sample	*Mean*	*Range*	*Source*
10	0·41	–	Captive in Europe? (113)
4	0·406	0·388–0·422	Captive at Slimbridge (C. Tyler pers. com.)

CLUTCH SIZE OF BIRDS AT SLIMBRIDGE

1. *Mean clutch size*

	No. of clutches	*No. of eggs*	*Mean*	*Range*
All clutches	674	2,686	3·99	1–6
Excluding those of one and two eggs	601	2,571	4·28	(3–6)

2. *Range of clutch sizes produced*

No. of eggs	1	2	3	4	5	6	*Total*
No. of clutches	31	42	87	230	247	37	674
%	4·6	6·2	12·9	34·1	36·7	5·5	100

Appendix 5: Gosling weights (g)

AT HATCHING (4–24 hours old)

	Sample	*Mean*	*Range*	*Standard error*
Slimbridge,	310	93·6	63·5–115·5	0·54
Pohakuloa (10)	87	101·9	90·6–114·4	–

DURING GROWTH

Slimbridge	*Sample*	*Mean*	*Range*	*Standard error*	
At hatching	310	93·6	63·5–115·5	0·54	
week					
1	56	198·3	135–280	4·48	‡
2	29	466·7	280–610	13·02	‡
3	18	772·8	600–890	18·47	‡
4	5	1,015·0	810–1,110	53·26	‡
5	5	1,168·0	980–1,240	47·89	NS
6	5	1,426·0	1,210–1,540	53·54	†
7	5	1,564·0	1,420–1,650	39·65	NS
8		no weights taken	–	–	
9	5	1,698·0	1,550–1,900	61·75	
10	5	1,767·0	1,660–2,000	65·18	NS

Note: from week 4 the sample consisted of 3 females and 2 males

KEY to Standard error.
NS means weight not significantly different.
† (P<0·01)
‡ (P<0·001) means weight very different.

Appendix 6: Nene breeding performance in captivity

Production of Nene at Slimbridge over 25 years: 1952–66 taken from Johnstone (72). 1967–72 data collected by Janet Kear, by kind permission of Mr S. T. Johnstone. All are descended from two females of Shipman origin, known as Emma & Kaiulani (Chapter 4).

Year	*Breeding females*	*No. of eggs*	*Infertile*	*Addled*	*Dead in shell*	*Hatched*	*Reared*	*% fertility*
1952	2	19	5	5	0	9	9	74
1953	2	17	10	2	0	5	4	41
1954	5	24	13	4	2	5	4	46
1955	6	31	22	3	0	6	4	29
1956	6	62	40	4	2	16	15	35
1957	7	70	42	11	3	14	5	40
1958	9	77	43	9	4	21	19	44
1959	15	91	32	18	6	24	21	65
1960	18	150	70	42	13	25	20	54
1961	21	140	76	25	10	29	26	46
1962	20	146	108	16	8	14	13	26
1963	18	120	83	11	6	20	16	31
1964	16	72	42	5	9	16	14	42
1965	16	80	47	11	0	22	17	42
1966	14	63	43	5	2	13	8	32
1967	22	123	46	25	12	40	35	57
1968	24	74	27	8	9	30	25	63
1969	15	77	11	18	14	34	32	86
1970	28	130	45	13	16	56	51	65
1971	29	132	34	21	19	58	54	74
1972	25	115	45	29	9	32	24	61
1973	25	118				47	42	
1974		218				69	57	
1975						70	60	
1976							72	
1977							83	
1978	25	120					35	
1979	30	169					17	
Total							782	

Dates of first eggs laid in captivity at Slimbridge and at Pohakuloa

	Slimbridge	Pohakuloa		Slimbridge	Pohakuloa
1951	2 Mar*	–	1966	9 Feb	4 Oct
1952	18 Feb	–	1967	11 Feb	22 Oct
1953	13 Feb	5 Dec	1968	16 Feb	11 Oct
1954	24 Feb	18 Nov	1969	30 Jan	30 Sept
1955	8 Feb	14 Nov	1970	3 Feb	17 Oct
1956	5 Feb	29 Oct	1971	3 Feb	10 Oct
1957	30 Jan	2 Nov	1972	8 Feb	12 Sept
1958	7 Feb	29 Oct	1973	10 Feb	3 Oct
1959	5 Feb	4 Nov	1974	26 Jan	30 Aug
1960	9 Feb	3 Nov	1975	7 Feb	13 Sept
1961	24 Jan	21 Sept	1976	12 Feb	15 Oct
1962	6 Feb	17 Nov	1977	5 Feb	26 Oct
1963	27 Feb	12 Nov	1978	9 Feb	15 Oct
1964	27 Jan	24 Oct	1979	8 Feb	31 Oct
1965	24 Feb	9 Nov			

* Two females present but no male.

Productivity of the Nene at Slimbridge 1951 to 1974

1: Production of eggs with age in 250 females

Age (years)	2	3	4	5	6	7 and 8	9 and 10	11+
No. of ♀	124	103	84	70	57	65	22	9
No. of laying ♀	72	92	74	63	52	58	17	4
No. of eggs laid	365	570	531	427	342	376	92	12
No. eggs per laying ♀	5·07	6·20	7·18	6·78	6·58	6·48	5·41	3·00
Range	0–13	0–15	0–14	0–16	0–16	0–15	0–11	0–8
Standard error	0·30	0·37	0·37	0·45	0·42	0·43	0·75	1·68
	*	NS	NS	NS	NS	NS	NS	
No. of eggs per available ♀	2·94	5·53	6·32	6·10	6·00	5·78	4·18	1·33
No. of clutches	99	140	127	99	84	92	26	5
Clutch size	3·68	4·07	4·18	4·31	4·07	4·09	3·54	2·40
No. of repeat clutches	27	48	53	36	32	34	9	1
No. of repeat clutches per laying ♀	0·38	0·52	0·72	0·57	0·62	0·59	0·53	0·25

2: Production of goslings with age in 103 females who produced at least one gosling during their life span.

Age (years)	2	3	4	5	6	7 and 8	9 and 10	11+
No. of ♀	98	98	91	76	58	74	32	11
No. of goslings	113	184	220	184	96	112	39	2
No. of goslings per ♀	1·15	1·88	2·42	2·42	1·66	1·51	1·22	0·18
Range	0–7	0–12	0–10	0–14	0–12	0–12	0–9	0–2
Standard error	0·18	0·23	0·25	0·36	0·31	0·26	0·42	0·18

* NS NS NS NS NS NS

3: Fertility of the male in 95 pairs of Nene

Age (years)	2	3	4	5	6	7 and 8	9 and 10	11+
No. of productive pairs	37	60	55	48	40	40	24	16
No. of unproductive pairs	21	8	4	4	1	6	5	4
No. of eggs	205	378	431	357	264	309	152	139
No. of fertile eggs	108	218	261	200	135	136	82	67
% fertility	52·7	57·7	60·6	56·0	51·1	44·0	53·9	48·2
Mean no. of fertile eggs per productive pair	2·91	3·63	4·74	4·16	3·37	3·40	3·41	4·18
Range	0–10	0–12	0–19	0–25	0–13	0–18	0–18	0–19
Standard error	0·47	0·44	0·51	0·67	0·50	0·68	0·97	1·41

NS NS NS NS NS NS NS

NS means weight not significantly different
* means weight significantly different but only slightly so.

Summary of production at Pohakuloa from 1953/1954 through the 1971/1972 breeding season

Season	1953–1954	1954–1955	1955–1956	1956–1957	1957–1958	1958–1959	1959–1960	1960–1961	1961–1962	1962–1963	1963–1964	1964–1965	1965–1966	1966–1967	1967–1968	1968–1969	1969–1970	1970–1971	1971–1972
No. breeding pairs	6	4	5	6	6	15	14	15	16	16	17	17	20	24	30	40	30	29	26
No. producing	4	4	5	6	6	15	14	15	16	16	16	16	20	23	27	38	30	29	21
First clutch	4	4	5	6	6	15	14	15	16	16	16	16	20	23	27	38	30	29	21
Second clutch	0	1	1	2	5	12	13	14	14	16	15	15	16	19	18	26	13	25	18
Third clutch	0	0	1	2	4	6	8	10	9	13	13	8	11	7	2	0	1	5	2
Fourth clutch	0	0	0	1	2	3	0	0	2	1	1	0	0	0	0	0	0	0	0
Total eggs	14	16	27	46	76	118	149	174	181	204	202	176	197	208	196	259	180	260	185
Eggs per clutch	3·5	3·2	3·8	4·1	4·4	3·3	4·3	4·4	4·4	4·4	4·5	4·5	4·2	4·2	4·2	4·0	4·1	4·5	4·5
Eggs per goose	3·5	4·0	5.4	7·7	12·6	7·9	10·6	11·6	11·3	12·7	12·6	11·0	9·9	9·1	7·3	6·8	6·0	8·6	8·9
No. fertile, LG*	6	4	18	22	15	28	56	97	87	130	106	121	138	143	151	200	145	191	139
No. fertile, DG†	0	1	0	0	1	4	5	18	18	17	28	22	18	10	6	7	12	28	4
No. infertile	8	10	8	16	47	76	86	49	68	50	56	26	32	43	26	36	19	29	29
No. damaged[1]	0	1	1	8	13	10	2	10	8	7	11	7	9	12	13	16	10	12	13
% eggs with fertile, LG	42·8	35·0	66·7	47·8	19·7	23·7	37·5	55·7	48·1	63·7	52·7	68·7	70·1	68·7	77·0	77·2	80·6	73·5	75·0
% eggs fertile	42·8	31·2	61·7	47·8	21·1	27·1	40·9	66·2	58·1	72·7	66·3	81·3	79·2	73·7	80·2	80·0	88·3	84·2	77·0
No. hatched	4	4	8	14	8	11	27	40	52	65	47	50	81	93	121	176	122	145	111
% hatchability[2]	66·8	100·0	44·5	63·7	53·3	39·2	48·2	41·2	59·8	50·0	44·3	41·3	58·7	65·0	80·1	88·0	78·7	66·2	77·6
Mortality[3]	0	0	0	2	4	1	8	7	6	5	5	6	11	1	5	13	8	13	7
% mortality	0	0	0	14·3	50·0	9·1	29·6	17·5	11·5	7·8	10·6	12·0	13·6	1·1	4·1	7·4	6·5	8·9	5·0
Goslings per goose[4]	1·0	1·0	1·6	2·3	0·7	0·7	1·9	2·7	3·2	4·1	2·9	3·1	4·1	4·0	4·5	4·6	3·8	5·0	4·0

1 These are eggs broken in the nest; soft-shelled; abnormally small, etc. Fertility undetermined.
2 Percentage of eggs with fertile, live germs that were successfully hatched.
3 Only post-hatch mortality (occurring within the first two weeks) is included here.
4 This represents production per goose of goslings successfully hatched.
* LG means live germ at 10 days of incubation.
† DG means dead germ at 10 days of incubation.

Appendix 7: Mortality of Slimbridge Nene (over 25 years)

Post mortem findings of ringed adult Nene (more than 3 months old) of Slimbridge origin. The table includes 149 females and 120 males, but more than one condition may be found in an animal upon post mortem examination.

	♂	♀	*Total*
Fungal disease (Aspergillosis)	8	10	18
Bacterial diseases			
TB	12	23	35
Enteritis		7	7
Others	2		2
Parasitic conditions (11)			
Gizzard worm (*Amidostoma*)	32	42	74
Gape worm (*Cyathostoma*)	5	3	8
Tapeworms	1		1
Roundworms (*Heterakis*)	3	2	5
Respiratory mites	1	3	4
Mallophaga	1		1
Coccidosis	1		1
Others	2	2	4
Respiratory disorders			
Pneumonia	3	7	10
Pulmonary congestion	2	7	9
Degenerative disease			
Atherosclerosis	32	25	57
Amyloid disease	6	4	10
Kidney malfunction	9	6	15
Liver malfunction	5	1	6
Digestive disorders (gizzard erosion, etc.)	1	5	6
Reproductive disorders (egg peritonitis, etc.)		22	22
Cardiac disease	9	10	19
Avian Pox (83) Plate 23	12		12
Nervous disorders (132)	1	1	2
Tumour	1	1	2
Trauma	24	20	44

Mortality at Slimbridge of adult Nene month by month

	Males	*Females*
Jan	10	3
Feb	9	14
Mar	8	34
Apr	11	17
May	7	10
Jun	8	8
Jul	9	4
Aug	2	10
Sep	6	6
Oct	7	7
Nov	10	6
Dec	6	8
Totals	93	127

Mortality at Slimbridge of Nene older than three months

	Males	*Females*
Under 1 year	11	11
Years		
1–2	7	9
2–3	11	12
3–4	8	18
4–5	11	13
5–6	12	11
6–7	8	9
7–8	3	16
8–9	2	15
9–10	6	2
10–11	5	2
11–12	2	3
12–13	1	4
13–14	4	2
14–15	0	1
15–16	1	1
16–17	1	0
17–18	1	0
Totals	94	129

Appendix 8: Rainfall at Keauhou and Kahaku Sanctuaries

1: *Rainfall for Keauhou Sanctuary* (cm)

	1960	1961	1962	1963	1964
January	15·16	3·35	4·75	13·08	11·66
February	11·10	10·56	6·27	4·42	10·01
March	18·92	4·70	20·70	32·26	52·52
April	19·12	3·18	7·49	43·69	2·08
May	11·79	8·51	11·41	15·09	6.40
June	2·56	0·02	9·22	14·37	1·16
July	3·70	2·18	2·03	12·83	2·94
August	2·46	1·98	0·81	2·18	8·00
September	16·00	9·20	12·77	8·15	5·28
October	7·01	6·91	2·77	2·29	8·15
November	12·21	28·72	1·04	8·38	20·12
December	4·75	48·20	4·77	6·91	16·82
Total	124·78	127·51	84·03	163·65	145·14

	1965	1966	1967	1968	1969
January	8·40	4·29	7·67	28·07	37·19
February	18·31	18·31	11·15	14·20	35·12
March	12·55	7·18	29·57	21·34	10·31
April	16·28	16·49	36·63	35·20	14·24
May	32·49	10·54	18·11	12·01	0·91
June	1·93	0·40	23·13	6·60	2·84
July	4·97	10·92	21·90	12·57	7·24
August	3·15	8·40	21·99	5·79	12·29
September	13·69	11·81	3·73	17·76	4·77
October	11·18	13·31	8·76	4·37	2·64
November	22·91	37·54	36·14	Stolen	5·08
December	18·34	25·78	25·73	34·04	4·67
Total	164·20	164·97	244·51	191·95	137·30

Rain gauge records were from the National Park gauge at the end of the truck trail at 2,042 metres elevation adjacent to the Keauhou Sanctuary.

2: Rainfall records for Keauhou 2 Sanctuary area (cm)

	1966	1967	1968	1969
January	2·87	6·10	10·67	17·27
February	2·21		8·38	6·10
March	0·83	14·48		3·56
April	3·02	9·65	26·16	2·79
May	12·50	3·30	6·35	5·33
June	8·89		12·45	2·03
July		13·72	9·40	2·54
August		3·30	6·10	2·29
September		6·91		0·25
October	54·86	6·60	22·35	0·25
November	3·15	4·57	3·30	0·58
December	2·29	7·57	12·07	3·43
Total	90·62	76·20	117·23	46·42

Gauge located at Ahu Umi, 1,588 metres elevation.
Where no record occurs in the above table the gauge was not read.
Rainfall for those months is included in the next month's total.

3: Rainfall records for Kahuku Sanctuary area (cm)

	1956	1957	1958	1959	1960
January					
February					
March	102·11	56·13	41·66	71·63	41·15
April					
May					
June	81·79	64·52	61·21	43·69	51·05
July					
August					
September	61·47	106·93	91·69	97·54	61·47
October					
November					
December	86·87	71·63	28·96	40·74	43·69
Total	332·24	299·21	223·52	253·60	197·36

Rainfall records for Kahuku Sanctuary area —— contd.

	1961	1962	1963	1964	1965	1966
January						
February						
March	38·61	41·15	53·85	44·45	50·85	45·21
April						
May						
June	56·13	62·00	60·96	38·86	90·42	39·12
July						
August						
September	67·56	59·44	35·97	28·32	67·06	95·25
October						
November						
December	61·47	21·84	18·34	52·83	72·64	—
Total	223·77	184·43	169·12	164·46	280·97	179·58

Gauge located at Punaluu-Kahawai at 1,880 metres elevation.
This gauge was read on a quarterly basis.

Bibliography

1 Allen, G. A. 1976. Raising Nene Geese. *Game Breeders Gazette* 25: 17–19.

2 Allen, G. E. 1950. Nene – rare bird of Hawaii. *Nature Magazine*, November: 464–465.

3 Anon. 1833. (Exhibition of geese from the Sandwich Islands). *Proc. Zool. Soc. Lond.*: 65.

4 Anon. 1851. *Catalogue of the Menagerie and Aviary of Knowsley*. Liverpool: Printed by Joshua Walmsley, 50 Lord St.

5 Anon. 1932. The Hawaiian Goose. *Hawaiian Forester and Agriculturist* 29: 95.

6 Anon. 1951. The Ne-ne project. *Wildf. Trust Ann. Rep.* 3: 54–56.

7 Anon. 1955. The present status of the Ne-ne or Hawaiian Goose. *Wildf. Trust. Ann. Rep.* 7: 47–50.

8 Anon. 1963. Return of Slimbridge-reared Nene to Hawaii. *Wildf. Trust Ann. Rep.* 14: 17–18.

9 Anon. 1964. Ne-Ne progress report 1963. *Wildf. Trust Ann. Rep.* 15: 15–16.

10 Anon. 1972. A report of the Nene restoration program. State of Hawaii Dept. of Land & Nat. Reserves, Div. of Fish and Game.

11 Avery, R. A. 1966. Helminth parasites of wildfowl from Slimbridge, Glos. 1. Parasites of captive Anatidae. *J. Helminthology* 40: 269–280.

12 Baker, J. H. 1951. News of wildlife and conservation. *Audubon Mag.* 53, July–August: 255–256.

13 Baldwin, P. H. 1945. The Hawaiian goose, its distribution and reduction in numbers. *Condor* 47: 27–37.

14 Baldwin, P. H. 1947. Foods of the Hawaiian goose. *Condor* 49: 108–120.

15 Bellrose, F. C., Scott, T. G., Hawkins, A. S. & Low, J. B. 1961. Sex ratios and age ratios in North American ducks. *Bull. Ill. Nat. Hist. Serv.* 27: 385–474.

16 Berger, A. J. 1969. The breeding season of the Hawaii 'Amakihi. *Occ. Papers Bernice P. Bishop Mus.* 24: 1–8.

17 Berger, A. J. 1970. The present status of the birds of Hawaii. *Pacific Science* 24: 29–42.

18 Berger, A. J. 1972a. Hawaiian birds 1972. *Wilson Bull.* 84: 212–222.

19 Berger, A. J. 1972b. *Hawaiian Birdlife*. Honolulu: Hawaii U.P.

20 Berger, A. J. 1975. Hawaii's dubious distinction. *Defenders* 50: 491–496.

21 Berger, A. J. 1978. Reintroduction of Hawaiian Geese. *Endangered Birds: Management Techniques for Preserving Threatened Species.* P.339–344. Madison, Wisconsin: Wisconsin U.P.

22 Blaauw, F. E. 1904. On the breeding of some of the waterfowl at Gooilust in the year 1903. *Ibis* (8)4: 67–75.

23 Blackman, T. M. 1944. A rare goose. *Natural History*, November: 407.

24 Boddam-Wetham, J. W. 1876. *Pearls of the Pacific.* London: Hurst and Blackett.

25 Brambell, M. R. 1977. Reintroduction. *Int. Zoo Yb.* 17: 112–116.

26 Breese, P. L. 1957. Increasing the Nene in their native Hawaiian habitat. *Game Breeders Gazette* 6: 12–16, 52–56.

27 Breese, P. L. 1961. Progress report on the programme of increasing the Hawaiian Nene in the wild. *Game Breeders Gazette* 10: 12–17.

28 Breese, P. L. 1963. Rare species of wild geese. Hawaiian Nene Goose (*Branta sandvicensis*). *Game Breeders Gazette* 12: 39–41.

29 Brigham, W. T. 1899. Hawaiian feather work. *Memoirs Bernice P. Bishop Mus.* Vol. 1, no. 1. 81 pp.

30 Brigham, W. T. 1903. Supplementary notes to an essay on ancient Hawaiian feather work. *Memoirs Bernice P. Bishop Mus.* vol. 1, no. 5. 19 pp.

31 Brigham, W. T. 1909. The volcanoes of Kilaueau and Mauna Loa. *Memoirs Bernice P. Bishop Mus.* vol. 2, noo. 4. 222 pp.

32 Brigham, W. T. 1918. Additional notes on Hawaiian feather work. *Memoirs Bernice P. Bishop Mus.* vol. 7, no. 1. 69 pp.

33 Bryan, E. M. Jr. 1940. A summary of the Hawaiian birds. *Proc. Sixth Pacific Science Congress* 4: 185–189.

34 Bryan, L. W. 1947. Twenty-five years of forestry work of the island of Hawaii. *Hawaiian Planters Record* 51: 1–80.

35 Bryan, W. A. 1901. *A Key to the Birds of the Hawaiian Group.* Honolulu: Bishop Mus. Press.

36 Bryan, W. A. 1906. Egg of the Hawaiian Goose. *Occ. Papers Bernice P. Bishop Mus.* 2: 21–22.

37 Carpentier, J. 1964. Success d'elevage chez le Nenes ou Bernacles d'Hawaii et Roul-rouls. *Zoo Antwerp* 29: 110–112.

38 Delacour, J. 1949. The Hawaiian Goose, or Nene, in Europe. *Avic. Mag.* 55: 130–131.

39 Delacour, J. 1954. *The Waterfowl of the World.* Vol. 1. London: Country Life.

40 Delacour, J. & Mayr, E. 1945. The family Anatidae. *Wilson Bull.* 57: 3–55.

41 Dole, S. B. 1869. A synopsis of birds hitherto described from the Hawaiian Islands. *Proc. Boston Soc. Nat. Hist.* 12: 294–309.

42 Dole, S. B. 1879. List of birds of the Hawaiian Islands. *Thrum's Hawaiian Almanac and Annual 1879*: 41–58.

43 Dunmire, W. W. 1961. New hope for the Nene. *Nat. Parks Mag.* Aug. 1961: 4–7.

44 Elder, N. B. 1951. Can the Nene come back? *Audubon Mag.* 53, Jan.–Feb.: 24–30.

45 Elder, W. H. 1957a. Nene – Hawaii's official bird. *Paradise of the Pacific* 69, June: 12–15.

46 Elder, W. H. 1957b. Objectives of the Nene study. *Elepaio* 17: 47–48.

47 Elder, W. H. 1958a. A research report on the Hawaiian Goose. Intern. Comm. Bird. Pres. Pan-American Sect., Research Rep. No. 3: 1–8.

48 Elder, W. H. 1958b. Ne-ne in Hawaii. Preliminary report on the Ne-ne in Hawaii. *Wildfowl Trust Ann. Rep.* 9: 112–117.

49 Elder, W. H. & Woodside, D. H. 1958. Biology and management of the Hawaiian goose. *Trans. North American Wildlife Conf.* 23: 198–215.

50 Ellis, W. 1782. *An Authentic Narrative of a Voyage Performed by Captain Cook and Captain Clerke, etc.* London: G. Robinson.

51 Ellis, W. 1825. *A Journal of a Tour around Hawaii, the largest of the Sandwich Islands.* New York: Crocker & Brewster.

52 Ellis, W. 1917. *A Narrative of a Tour through Hawaii, or Owyhee, etc.* Honolulu: Hawaiian Gazette Co. Ltd. (a reprint of the 1827 London Edition).

53 Evans, M. E. 1979. A visit to the House of the Sun. *Wildfowl News* No. 80: 21–22.

54 Force, R. W. & M. 1968. *Art and Artifacts of the 18th Century.* Objects in the Leverian Mus. as painted by Sarah Stone. Honolulu: Bishop Mus. Press.

55 Frädrich, H. 1975. Die Gans de Lord Sandwich. *Der Zoofreund* (Hannover) 17: 10–11.

56 Gray, A. P. 1958. *Bird Hybrids.* London: Commonwealth Agricultural Bureaux.

57 Griswold, J. A. 1966. Persistence pays off, rare geese bred. *America's First Zoo* (Philadelphia) 18(2): 14.

58 Gaselee, J. 1963. The precious Nene. *Animals* 1: 15–18.

59 Gilbert, S. 1979. The Hawaiian goose or nene *Branta sandvicensis* breeding programme at the National Zoological Park, *Washington. Int. Zoo. Yb.* 19: 139–143.

60 Hachisuka, M. 1928. Variation among birds. Hawaiian × Chinese goose hybrid. *Tori* 5: 498.

61 Halliday, T. 1978. *Vanishing Birds: their Natural History and Conservation.* London: Sidgewick & Jackson.

62 Henshaw, H. W. 1902. *Complete list of the Birds of the Hawaiian Posessions, with Notes on their Habits.* Honolulu: Thos. G. Thrum.

63 Humphrey, P. S. 1958. The trachea of the Hawaiian Goose. *Condor* 60: 303–307.

64 Humphreys, P. N. 1972. Brief observations on the semen and spermatazoa of certain passerine and non-passerine birds. *J. Reprod. Fert.* 29: 327–336.

65 Hutton, F. W. 1871. *Catalogue of the Birds of New Zealand*. Geol. Survey of N.Z.

66 Johnsgard, P. A. 1965. *Handbook of Waterfowl Behavior*. Ithaca: Cornell UP.

67 Johnsgard, P. A. 1968. *Waterfowl: their Biology and Natural History*. Lincoln: Nebraska UP.

68 Johnsgard, P. A. 1971. Observations on sound production in the Anatidae. *Wildfowl* 22: 46–59.

69 Johnston, V. R. 1962. Ke Kua'aina of Hawaii. *Sierra Club Bull.*, Oct. 1962: 4–5.

70 Johnstone, S. T. 1959. Notes from the Wildfowl Trust. *Avic Mag.* 65: 37–39.

71 Johnstone, S. T. 1961. The breeding season 1960. *Wildfowl Trust Ann. Rep.* 12: 7–9.

72 Johnstone, S. T. 1967. Breeding Hawaiian Geese. *Avic. Mag.* 73: 86–88.

73 Johnstone, S. T. 1969. Ein Vogel wird vordem Aussterben gerettet. *Vogel-Kosmos* 6: 15–17.

74 Kear, J. 1964. Colour preference in young Anatidae. *Ibis* 106: 361–369.

75 Kear, J. 1965. The internal food reserves of hatching Mallard ducklings. *J. Wildl. Mgmt.* 29: 523–528.

76 Kear, J. 1968. The calls of very young Anatidae. *Beihefte der Vogelwelt* 1: 93–133.

77 Kear, J. 1969. Hawaiian Goose or Nene. *Birds of the World* 1: 218–20.

78 Kear, J. 1973. Notes on the nutrition of young waterfowl, with special reference to slipped-wing. *Int. Zoo Yb.* 13: 97–100.

79 Kear, J. 1975a. Breeding of endangered wildfowl as an aid to their survival. *Breeding Endangered Species in Captivity*. Pp. 49–60. London: Academic Press.

80 Kear, J. 1975b. Returning the Hawaiian Goose to the wild. *Breeding Endangered Species in Captivity*. Pp. 115–124. London: Academic Press.

81 Kear, J. 1977. The problems of breeding endangered species in captivity. *Int. Zoo Yb.* 17: 5–14.

82 Kear, J. 1978. Captive propagation of waterfowl. *Endangered Birds: Management Techniques for Preserving Threatened Species*. Pp. 243–249. Madison, Wisconsin: Wisconsin U.P.

83 Kear, J. & Brown, M. 1976. A pox-like condition in the Hawaiian Goose. *Int. Zoo Yb.* 16: 133–134.

84 Kear, J. & Murton, R. K. 1973. The systematic status of the Cape Barren Goose as judged by its photo response. *Wildfowl* 24: 141–143.

85 Kramer, R. J. 1971. *Hawaiian Land Mammals*. Rutland, Vermont: Charles E. Tuttle Co.

86 Kuhme, W. 1974. Problemzuchten im Kölner Zoo: Hawaiigans (*Branta Sandvicensis*). *Zeitschrift des Kölner Zoo* 17 (3): 97–102.

87 Lack, D. 1968a. *Ecological Adaptations for Breeding in Birds*. London: Methuen.

88 Lack, D. 1968b. The proportion of yolk in the eggs of waterfowl. *Wildfowl* 19: 67–69.

89 Lecomte, J. 1954. A propos de l'oie d'Hawai. *Terre et la Vie* 101: 151–152.

90 Lint, K. C. 1962. The Nene – State bird of Hawaii. *Zoonooz.* San Diego zool. Soc., Feb. 1962: 10–14.

91 Locey, F. H. 1937. Introduced game birds of Hawaii. *Paradise of the Pacific* 49: 5–6, 27–30.

92 Lubbock, M. 1975. The Ne-ne at home. *Wildfowl Trust Bull.* No. 72: 11.

93 MacFarlane, J. R. H. 1887. Notes on birds in the Western Pacific, made in HMS 'Constance', 1883–1885. *Ibis* (5)5: 201–215.

94 Malo, D. 1951. Hawaiian Antiquities (Moolelo Hawaii). Trans from the Hawaiian by Nathanuel B. Emerson. *Bernice P. Bishop Mus. Spec. Publication* 2. 2nd ed.

95 Matthews, G. V. T. 1973. Some problems facing captive breeding and restoration problems for waterfowl. *Int. Zoo. Yb.* 13: 8–11.

96 Miller, A. H. 1937. Structural modifications in the Hawaiian Goose (*Nesochen sandvicensis*), a study in adaptive evolution. *Univ. Calif. Publ. Zool.* 42, No. 1. 79 pp.

97 Mills, S. 1978. What's wrong with the Nene in Hawaii? *Oryx* 14: 359–361.

98 Montlesum, A. 1886. Note sur les Palmipedes Lamellirostres. *Bull. Soc. Acc.*: 132–167.

99 Mountfort, G. 1978. *Back from the Brink.* London: Hutchinson.

100 Munro, G. C. 1944. *Birds of Hawaii.* Honolulu: Tongg Publ. Co.

101 Murton, R. K. & Kear, J. 1973. The nature and evolution of the photoperiodic control of reproduction in certain wildfowl. *J. Reprod. Fert. Suppl.* 19: 67–84.

102 Nisbet, I. C. T. 1976. Pacific follies or the ravishing of Hawaii. *Tech. Rev.* 78(6): 8–9.

103 Peale, T. R. 1848. United States Exploring Expedition, 1838–1842. *Mammalia and Ornithology*, vol. 8. Philadelphia: Printed by Sherman.

104 Perkins, R. C. L. 1903. *Fauna Hawaiiensis: Vertebrata* (*Aves*). In Zoology of the Sandwich (Hawaiian) Islands. David Sharp, editor, vol. 1, part IV, pp. 368–465.

105 Pratt, J. J. 1970. The Nene's return to Maui. *Modern Game Breeding* 6 (4): 10–12.

106 Pratt, J. J. 1971. A breakthrough for the Maui Nene. *Elepaio* 32: 3.

107 Pratt, J. J. 1972a. Hawaiian Geese. *Elepaio* 33: 1.

108 Pratt, J. J. 1972b. Research study proposal for investigation of behaviour of the Hawaiian Goose under the 'Nene Park' plan. *Elepaio* 33: 33–34.

109 Ripley, S. D. 1958. The Nene can be saved. *Animal Kingdom*, June: 82–89.

110 Ripley, S. D. 1965. Saving the Nene, world's rarest goose. *National Geographic* 128 (5) November: 744–754.

111 Rothschild, W. 1893–1900. *The Avifauna of Laysan and the Neighbouring Islands*. London: R. H. Porter.

112 Salvadori, T. 1895. *Catalogue of the Birds in the Collection of the British Museum*. Vol. 27. London: Taylor & Francis.

113 Schönwetter, M. 1960–61. *Handbuch der Oologie*. Ed. W. Meise. Berlin: Akademieverlag.

114 Schwartz, C. W. & Schwartz, E. R. 1949. *A Reconnaissance of the Game Birds of Hawaii*. Honolulu Board of Commissioners of Agriculture and Forestry.

115 Sclater, P. L. 1880. List of the certainly known species of Anatidae. *Proc. zool. Soc. Lond:* 496–536.

116 Scott, P. 1956. Nenes at the Wildfowl Trust in England. *Elepaio* 16: 66–67.

117 Scott, P. 1961. *The Eye of the Wind*. London: Hodder & Stoughton.

118 Scott, P. 1962. A project for a Nene park in Hawaii. *Elepaio* 22: 11.

119 Sekora, P. O. 1975. Refuge for rare species. *Defenders* 50: 506–511.

120 Smith, J. D. 1952. The Hawaiian goose (Nene) restoration program. *J. Wildl. Mgmt.* 16: 1–9.

121 Stanley, E. S. (13th Earl of Derby). 1834. A note on a specimen of a young Sandwich Island goose. *London and Edinburgh Philosophical Magazine and Journal of Science*, 5: 233–235; and *Proc. Zool. Soc. Lond.* Part II: 41–43.

122 Stearns, H. T. 1973. Geologic setting of the fossil goose bones found on Molokai Island, Hawaii. *Occ. Papers Bernice P. Bishop Mus.* 24: 156–163.

123 Titcomb, M. 1956. The Nene. *Elepaio* 16: 63–65.

124 Tomich, P. Q. 1969. Mammals in Hawaii. *Bernice P. Bishop Mus. Spec. Publ.* 57. Honolulu. 238 pp.

125 Trumbull, G. 1976. The return of Hawaii's wild goose. *Nat. Parks & Conserv. Mag.* 50: 16–19.

126 Van Tyne, J., & Berger, A. J. 1976. *Fundamentals of Ornithology*. New York: John Wiley & Sons.

127 Vigors, N. A. 1833. List of the animals in the gardens. In: Reports of the Council & Auditors of the Zoological Society of London.

128 Walker, R. L. 1967. A brief history of exotic game birds and mammal introductions into Hawaii, with a look to the future. Conference of Western Assoc. of State Game and Fish Commissioners, Honolulu, July 19, 1967.

129 Walker, R. L. 1970. Nene restoration project report. *Elepaio* 31: 1–7.

130 Warner, R. E. 1968. The role of introduced diseases in the extinction of the endemic Hawaiian avifauna. *Condor* 70: 101–120.

131 Wetmore, A. 1943. An extinct goose from the island of Hawaii. *Condor* 45: 146–148.

132 Wight, P. A. L. 1976. The histopathology of a cerebral lipidosis in the Hawaiian Goose, *Branta sandvicensis*. *Neuropath. & Applied Neurobiol.* 2: 335–347.

133 Wilson, S. B. & Evans, A. H. 1890–1899. *Aves Hawaiiensis: The Birds of the Sandwich Islands*. London: R. H. Porter.

134 Woolfenden, G. E. 1961. Post cranial osteology of the waterfowl. *Bull. Florida St. Museum Biol. Sci.* 6: 1–129.

135 Woodside, D. H. 1956. Wild Nene on Hawaii. *Elepaio* 16: 67–68.

136 Woodside, D. H. 1961. Future for a state bird. *Pacific Discovery* 14: 24–26.

137 Woodside, D. H. 1975. Reprieve for the Nene. *Defenders* 50: 480–481.

138 Woodworth, J. R. 1956. A brief history of the Nene restoration project at Pohakuloa, Hawaii. *Elepaio* 17: 7–10.

139 Yealland, J. J. 1951. Notes on some birds of Hawaii. *Avic. Mag.* 57: 39–46.

140 Yealland, J. J. 1956. Hawaiian waterfowl. *Zoo Life* 11: 47–50.

141 Yocum, C. F. 1967. Ecology of feral goats in Haleakala National Park, Maui, Hawaii. *Amer. Midl. Nat.* 77: 418–451.

142 Zimmerman, D. R. 1974. Return of the Nene – A survival story in search of a happy ending. *Natural History* 83: 22–28.

143 Zimmerman, D. R. 1975. *To Save a Bird in Peril.* New York: Coward, McCann & Geoghegan, Inc.

An Hawaiian Bird Bibliography in three parts was published in October 1978 by Winston E. Banko. This has been produced by the University of Hawaii at Manoa, Department of Botany.

Index

Anas platyrhynchos laysanensis, 20, 125, 128
P. wyvilliana, 21, 125
Anser, 35
cygnoides, 60, 87
sandvicensis, 59
Asio flammeus sandwichensis, 57, 125
Atherosclerosis, 86, 94–96, 138
Avian pox, 42, 86, 119–120, 138, *Plate 23*

Baldwin, P. H., 24, 26–29, 41–42, 46, 69, 106, 143
Bands (see under Rings)
Banko, W. E., 13, 122, 149
Bernicla sandvicensis, 21, 59
Bison bonasus, 115
Bison, European, 115
Blaauw, F. E., 61–62, 95, 144
Blackman, T. M., 33–34, 144
Boddam–Whetham, J. W., 26, 144
Branta, 35
canadensis (see also Goose, Canada), 23
sandvicensis (see also Nene), 20, 59–60, 125
Breese, P. L., 69, 73, 74, 104, 116, 144
Buteo solitarius, 57, 125
Brigham, W. T., 23, 26, 28, 144
Bryan, W. A., 23, 32, 50, 132, 144

Cat, 31, 33, 109
Cereopsis novaehollandiae, 23, 61, 87
Chelonia mydas, 20
Conuropsis carolinensis, 61
Cook, Capt. James, 17, 20, 27, 33, 63, 80, 101
Culex pipiens, 120

Deer, Père David's, 115
Delacour, Dr Jean, 48, 50, 59, 62, 69, 132, 144
Derby, Earl of (E. S. Stanley), 51, 59–61, 148
Diseases, atherosclerosis, 86, 94–96, 138
bacterial, 77, 86, 92, 138
cardiac, 138
digestive, 138
fungal, 82, 86, 138
kidney, 138
liver, 77; 138
parasitic, 110, 138
pox, 42, 86, 119–120, 138
respiratory, 86, 138
reproductive, 77, 81, 98, 138
Display, 'forward threat' posture, 48–49 *Plate 7*
Dog, 31, 33, 45, 57, 102, 106, 109, 116
Dole, S. B., 62, 144
Down, cottony, 52, 116–118, *Plate 22*
hairy, 52
Duck, Hawaiian, 21, 121, 125
Laysan, 20, 125, 128

Ectopistes migratorius, 61
Elder, Dr W. H., 49, 54, 56–57, 73–74, 100, 104–105, 115, 116, 145
Ellis, W. (Surgeon), 27–28, 145
(Missionary), 28, 145

Elaphurus davidianus, 115
Emma, 80, 117, 134
Equus przewalskii, 115

Gadow, H., 62
Geochen rhuax, 23
Goat, 31, 32, 44, 102, 103, 116
Goose, Barnacle, 59, *Plate 3*
  Black Brant, *Plate 3*
  Canada, 23, 25, 35, 36, 40, 48, 52, 56, 86, *Plates 3, 13*
  Cape Barren, 23, 87
  Chinese, 28, 87
  fossil, 21–23
  Greylag, 86
  Hawaiian (see under Nene)
  Redbreasted, *Plate 3*
Grouse, Red, 119

Haleakala Crater, 23, 109–113, 122
Hawaiian birds, endemic, 125–126
  extinct, 33, 63, 119, 127
  origin of, 20–22
  rare and endangered, 127–128
Hawk, Hawaiian, 57, 106, 125
Henshaw, H. W., 26, 33, 50, 63, 88, 132, 145
Honeycreeper, Hawaiian, 21, 22, 126–128
Horse, Przewalski's, 115
Hualalai Mountain, 28, 30, 64, 66, 102, *Plate 10*
Humphrey, P. S., 36, 145

Io, 57, 106, 125
Island, Hawaii, 17, 19, 23, 27, 29, 41, 62–77, 100–109, 120, 121
  Kauai, 19, 20, 23
  Kure, 17, 19, 20
  Lanai, 19, 64
  Laysan, 18, 19, 20
  Leeward, 20
  Maui, 19, 23, 24, 41, 109–113, 120–121
  Midway, 17, 18, 19
  Molokai, 19, 23, 64
  Nihoa, 18, 19, 20
  Niihau, 19, 20, 23
  Oahu, 17, 19, 63, 64, 121

Johnsgard, Dr P. A., 36, 48–49, 86, 146
Johnstone, S. T., 54, 82, 134, 146

Kahili, 26
Kaiulani, 80, 81, 134
Kamehameha, 32, 80–81, 134
Keaau, 39, 57, 67, *Plate 1*
Kipuka, 31, 32, 41, 48, 100, 102, *Plate 4*
Koloa, 21, 121, 125
Kosaka, E., *Plate 19*

Lack, Dr David, 50–52, 54, 146–147
Lagopus lagopus, 119
Lee, Ah Fat, 13, 15, 118, *Plate 9*
Leucopsar rothschildi, 119
Locey, F. H., 65–66, 147

Malo, D., 24, 147
Matthews, Dr G. V. T., 116–122, 147
Mauna Kea, 32, 33, 56, 57, 103, *Plate 8*
Mauna Loa, 28, 31, 41, 49, 56, 66, 73, 74, 76, 103, 107, *Plates 4, 5, 18*
Meleagris gallopavo, 102
Miller, A. H., 39–41, 56, 147
Monachus schauinslandi, 20
Mongoose, 33, 105, 106, 109, 113, 116
Mosquito, 41, 86, 118, 120
Munro, G. C., 62–63, 147
Museum, Bernice P. Bishop, 23, 24, 50, 73, 87
  Merseyside County, Liverpool, 21, 60

National Park Programmes, Hawaii, 112, 122
  Maui, 113, 122
Nene, adaptations, 38–41
  breeding season, in England, 87–89, 93, 96, 135
    in Hawaii, 48, 76, 87–88, 96, 135
  causes for decline, 31–34
  clutch size, 50, 77, 89, 92, 132, 135, 137
  copulation, 49, 87
  description, 35, 130
  diet, 69, 93, 99, 110
  displays, 48–49
  eggs, 50, 60, 83, 89–92, 96, 131–132
  fertility, 68, 91–92, 116, 134, 136
  flight, 41, 94–95
  flocking, 56

food, 42–47, 93, 94, 99
foot, 40, 41
foster mothers, 72, 93, 120
genetic strain, 73, 93, 116–118
gosling, description, 52, 56
growth of, 95, 97
weight of, 52, 93, 133
habitat of, 41
hatchability, 68, 92–93
hunting season, 33
husbandry, 76–77, 81–83, 120
hybridisation, 87
incubation, 49–52, 98–99
longevity, 77, 86, 93
mortality, 54, 84–87, 98, 138–139
moult, 37–38, 54, 87, 98, *Plates 8, 14, 20, 21*
nest sites, in captivity, 76, 81–83
in the wild, 49
number of clutches, 51–52, 68, 72, 135, 137
origin, 22, 23
parasites of, 86, 110, 138
predators, 33, 42, 45, 54, 57, 103, 106, 112, 121
preen gland, 81, *Plate 21*
productivity, 70–71, 75, 81, 83, 134–137
range, 41
reproduction, 48–52, 77, 87
sex ratio, 85
skull, 39
sperm, 87, 88, 92
time of laying, 76, 83, 89, 98, 135
trachea, 36, 37
type specimen, 21
voice, 35–36, 53–54, 56
weight, 33, 36–38, 97–98, 129
Nesochen sandvicensis, 23, 59, 61
Newton, A., 62

Owl, Hawaiian Short-eared, 57, 125

Palmer, H., 62–63
Parakeet, Carolina, 61
Patuxent, Maryland, 96, 115
Peale, T. R., 21, 24, 147
Perkins, R. C. L., 23, 63, 147
Pig, 31–33, 44, 102, 103, 106
Pigeon, Passenger, 61
Pohakuloa, 15, 66, 68, 70–77, 96–99, 117, 118, 120, 129, 131, 135, 137, *Plates 8, 9, 15*
Pratt, J. J., 24, 41, 112, 147
Propagation of Nene, at Pohakuloa, 66–69, 70–71, 73
at Slimbridge, 81–96
Pueo, 57, 125
Puuwaawaa, 28, 30, 63, 101

Rats, 31, 33, 106
Release pens, on Hawaii, 100–103, 113
on Maui, 109–111, 113
problems in, 105–106
Rings (Bands), 85, 105, 107, 130, *Plate 5*
Ripley, Dr S. D., 69, 110, 111, 147
Rothschild, Lord, 62–63, 148

Salvadori, T., 59, 148
Sanctuaries, Kahuku, 70, 101, 102, 141–142
Keauhou 1, 49, 70, 88, 100, 101, 140, *Plates 4, 5, 10, 11*
Keauhou 2, 70, 102, 141, *Plates 10, 18, 19*
Kipuka-Ainahou, 56, 70, 103
Paliku, 70, 109
Sandwich, Earl of, 17
Schwartz, C. W. & E. R., 29–30, 66–67, 69, 148
Scott, P. M. (Sir), 15, 69, 79–80, 112–113, 122, 148
Seal, Hawaiian Monk, 20
Shipman, Herbert, 30, 56, 63, 67–68, 73, 80, *Plate 1*
Slimbridge (see under Wildfowl Trust)
Slipped wing, 99
Smith, J. D., 30–31, 67, 69, 73, 75, 79, 115, 146
Sonagram, 36, 53–54
Starling, Common, 119
Rothschild's, 119
Sturnus vulgaris, 119
Syrmaticus reevesi, 59

Turkey, 31, 102
Turtle, Green Sea, 20

'Upright threat' posture, *Plate 6*

Vigors, N. A., 21, 59, 148
Voice, of adult, 35–36
  of gosling, 53–56
Volcanoes National Park, 32, 56, 100, 112, 113, 122

Wetmore, A., 21–23, 148
'Wigwam' nest shelter, *Plate 12*
Wildfowl Trust, 15, 79–99, 115, 120, 129–136, 138–139, *Plates 2, 6, 12, 14, 16, 17, frontispiece*
Williams, J., 84
Wilson, S. B., 23, 26, 62–63, 148
Woodside, D. H., 13, 49, 56–57, 100, 107, 117, 145, 149, *Plates 10, 19*
Woodworth, J. R., 67, 74, 149

Yealland, J. J., 54, 69, 80, 149

Zoological Society of London, 21, 59, 61, 118
Zoo, Honolulu, 64, 67, 68, 73, 109, 120
  Washington, 119–120

WILL YOU HELP US ENSURE THE SURVIVAL OF A SPECIES?

In 1947, there were less than 50 Hawaiian Geese left in the world and the species was in acute danger of extinction. In this book you have read a detailed account of a success story: the Wildfowl Trust and others have ensured that, by captive breeding, the bird has increased to a point where it can again be established in the wild.

There are other waterfowl that are threatened; every year that passes adds new species to the list and brings those already endangered closer to the brink of extinction.

The Wildfowl Trust is at work saving habitat, influencing governments, researching into the biology of wild and captive waterfowl. Funds are urgently needed to support the Trust's efforts and to enable increasingly effective conservation measures to be undertaken.

EACH NEW MEMBER AND EVERY DONATION OR BEQUEST WILL HELP

The Wildfowl Trust is a Registered Charity and makes no profit. Membership is open to all and we shall welcome your support. Donations may be sent to and a suggested form of bequest or membership application form obtained from:

The Controller
The Wildfowl Trust
Slimbridge
Gloucester
GL2 7BT

Your assistance will provide a stronger voice with which to present the case for conservation, the moral support that sustains our efforts and the financial support essential to the continuity of our work.